Structural and Mechanical Behavior of Glassy Polymers

Structural and Mechanical Behavior of Glassy Polymers

M.S. Arzhakov, S.A. Arzhakov, and
G.E. Zaikov (Editors)

Nova Science Publishers, Inc.
Commack New York

Art Director: Maria Ester Hawrys
Assistant Director: Elenor Kallberg
Graphics: Susan A. Boriotti and Frank Grucci
Manuscript Coordinator: Phyllis Gaynor
Book Production: Gavin Aghamore, Joanne Bennette, Michelle Keller, Christine Mathosian and Tammy Sauter
Circulation: Iyatunde Abdullah, Cathy DeGregory and Annette Hellinger

Library of Congress Cataloging–in–Publication Data
available upon request

ISBN 1-56072-434-X

6080 Jericho Turnpike, Suite 207
Commack, New York 11725
Tele. 516-499-3103 Fax 516-499-3146
E-Mail: Novascience@earthlink.net

Printed in the United States of America

Contents

PREFACE

One of the fundamental problems of modern polymer science concerns finding correlation between structure and properties of polymers. As compared with semicrystalline polymers, this problem seems to be far more difficult for amorphous glassy polymers because no efficient experimental methods for studying their structure are available. In many cases, advances in this particular area are provided by examination of their physical and mechanical behavior.

In early 60s, structural and mechanical studies in polymer glasses were started under supervision by Prof. V.A. Kargin in Institute of Chemistry and Technology of Polymers (Dzerzhinsk, Russia) headed by Prof. S.A. Arzhakov. To approach this challenging problem, a method of temperature-induced relaxation of residual plastic deformation of glassy polymers was advanced. For the first time, temperature-induced elastic relaxation of residual plastic deformation in polymer glasses was shown to involve low-temperature and high-temperature components. High-temperature component of elastic relaxation of residual deformation is observed in a rather narrow temperature region near glass transition temperature. Evidently, the origin of this mode of relaxation was rationalized by the transition of the excited chain conformations to their initial equilibrium state provided by segmental mobility in polymer body. On contrary, low-temperature component of elastic relaxation of residual deformation takes place in a wide temperature range at temperatures well below glass transition temperature, i.e., within glassy state of polymer. Let us emphasize, that, at each temperature, low-temperature elastic relaxation is characterized by a certain equilibrium portion of recovery of residual deformation. This phenomenon was rationalized by temperature-dependent elasticity of glassy polymers as distinguished from a universal temperature-independent Hookean elasticity.

To clarify the mechanism of the low-temperature component of relaxation, the ideas concerning structural organization of glassy polymers were invoked. These speculations allowed one to appreciate a decisive role of supermolecular structures in polymer glasses for the development of their physical and mechanical properties. This concept was discussed at length in *S.A. Arzhakov, N.F. Bakeev and V.A. Kabanov, Vysokomol. Soedin., A, 15, 1154 (1973).*

In recent years, further activities in this direction were performed at Polymer Department (Moscow State University, Moscow, Russia) headed by Prof. V.A. Kabanov. In this collection of papers the results of these works were generalized taking into account the current advances in studying the structure of glassy polymers.

A complicated character of temperature-induced relaxation in the plastically deformed polymer glasses still attracts much attention of scientists. Modern approaches and theories concerning the origin of this phenomenon are discussed in *M.S. Arzhakov "Temperature-Induced Relaxation in Deformed Polymer Glasses"*. Consideration of the structure of glassy polymers as a set of structural sublevels with different packing densities allowed us to explane a two-stage character of temperature-induced relaxation of plastic deformation in polymer glasses and to advance a structural pattern for plastic deformation in polymer glasses, which is based on gradual involvement of the different structural sublevels in deformation and relaxation processes. To support this concept, the structural rearrangements in glassy polymers induced by plastic deformation were studied using positron annihilation lifetime spectroscopy (*I.B. Kevdina, M.S. Arzhakov and V.P. Shantarovich "Studies in Deformed Poly(methyl methacrylate) Using Positron Annihilation Lifetime Spectroscopy"*). The effect of the changes in initial structure on both mechanical and relaxation properties of glassy polymer is discussed in *M.S. Arzhakov, S.A. Arzhakov, V.V. Gustov, I.B. Kevdina and V.P.*

Shantarovich "Structural and Mechanical Behavior of Poly(methyl methacrylate) Containing Diethyl Siloxane Oligomer".

For polymer samples prepared from powderlike glassy polymers under the parallel action of pressure and temperature, studying temperature-induced relaxation provided a detailed information concerning the mechanism of monolithization of glassy thermoplastic powders. Moreover, this experimental evidence allowed us to clarify the effect of inner stresses stored in polymer bodies during deformation on the development of low-temperature component of temperature-induced relaxation. The generalized data are discussed in *M.S. Arzhakov and S.A. Arzhakov "Physical and Mechanical Behavior of Polymer Glasses. I. Polymeric Powders".*

For bulk polymer glasses, the comparison of physical and mechanical behavior under uniaxial and triaxial compression provided a support for structural pattern of plastic deformation in glassy polymers (*M.S. Arzhakov and S.A. Arzhakov "Physical and Mechanical Behavior of Polymer Glasses. II. Poly(methyl methacrylate)"*).

Generally speaking, physical and mechanical properties of polymer glasses are likely to be controlled by the co-existence of low-temperature and high-temperature components of relaxation. From this standpoint, the ratio between the above components of relaxation was treated as a new physical parameter describing not only plastic deformation of glassy polymers but, in general, the development of their physical behavior and mechanical response. These problems are discussed in papers *M.S. Arzhakov, S.A. Arzhakov, M.B. Lachinov, E.V. Nechaeva and D.V. Trachenko "Physical and Mechanical Behavior of Polymer Glasses. III. Copolymers of Methyl Methacrylate"* and *M.S. Arzhakov and S.A. Arzhakov "Physical and Mechanical Behavior of Polymer Glasses. IV. Cross-Linked Poly (methyl methacryalate)*, and summarized in review article *M.S.*

Arzhakov and S.A. Arzhakov "Physical and Mechanical Behavior of Polymer Glasses in Terms of Temperature-Induced Relaxation of Plastic Deformation".

In connection with this, the problems of modification of polymer glasses should be treated as controlled variations in the ratio between low-temperature and high-temperature components of relaxation (*M.S. Arzhakov, S.A. Arzhakov and V.A. Chernavin "A New Approach for the Preparation of Modern Polymeric Materials Based on Polymer Glasses"*). Let us emphasize, that considering of plastic deformation and its further temperature-induced relaxation in terms of the structural inhomogeneity of polymer glasses made it possible to develop a new technology for the preparation of orientation-strengthened shrinkproof polymeric materials based on poly(methyl methacrylate)

The ideas concerning the effect of structural inhomogeneity of glassy polymers on their physical and mechanical properties may be applied not only for fully amorphous polymer glasses but also to semicrystalline polymers. In many cases, physical and mechanical behavior of semicrystalline polymers is known to be controlled by their amorphous phase. Consideration of structural inhomogeneity of amorphous regions in semicrystalline polymers allowed one to understand the specific features of interaction of these materials with modifying low-molecular-mass agents (*M.S. Arzhakov "Structural Aspects of Modification of Polymers with Low-molecular-mass Substances"* and *M.S. Arzhakov, E.V. Nechaeva, N.V. Red'ko, A.V. Volkov, A.L. Kazarin, A.L. Volynskii and N.F. Bakeev "Interaction of Polyamide-6 with Mixtures of Oxyaromatic Compounds"*). This approach allows a selective modification of different structural levels within amorphous phase of semicrystalline polymers and, as a result, makes it possible to control physical and mechanical behavior of these materials. For model polymer systems based on polyamide-6, this problem is discussed in *M.S. Arzhakov, E.V. Nechaeva, I.I. Khairullin, A.V. Volkov, A.L. Volynskii and N.F. Bakeev "Physical and Mechanical Behavior of Polyamide-6 Containing Oxyaromatic Compounds".*

Hence, interpretation of temperature-induced relaxation of residual plastic deformation of polymer glasses in terms of their structural inhomogeneity allowed us to advance the structural pattern for plastic deformation and to describe the development of physical and mechanical properties of these polymer materials. On the basis of this approach, scientific and technological principles for controlled modification of polymer plastics were worked out. This scientific direction may be referred to as "structural mechanics of glassy polymers". We are well aware, that some assumptions of the concept proposed invite further discussion. We would like to emphasize, that the principal objective of this work was far from providing a refined theory but to attract the attention of scientists to the fascinating problem concerning the structural pattern of mechanical behavior of glassy polymers.

Contributors:

M.S. Arzhakov

S.A. Arzhakov

N.F. Bakeev

V.A. Chernavin

V.V. Gustov

L.A. Kazarin

I.B. Kevdina

I.I. Khairullin

M.B. Lachinov

E.V. Nechaeva

N.V. Red'ko

V.P. Shantarovich

D.V. Trachenko

A.V. Volkov

A.L. Volynskii

Temperature-Induced Relaxation in Deformed Polymer Glasses

M.S. Arzhakov

Polymer Department, Faculty of Chemistry, Lomonosov Moscow State University

Vorob'evy Gory, Moscow, 119899 Russia

Temperature-induced relaxation of polymer glasses after their plastic deformation involves recovery of residual deformation and relaxation of inner energy stored by polymer during deformation. In general case, this process is realized in two temperature regions: at temperatures below polymer glass transition point and in temperature region of glass transition. Complex behavior of temperature-induced relaxation is considered in terms of specific features of elementary relaxation transitions and conformational changes in the deformed polymer as well as according to the speculations concerning supermolecular organization of polymer glasses. Principal approaches to the description of this phenomenon based on the examination of literature data are outlined. The importance of the research activities focused on temperature-induced relaxation of the deformed polymers is related to the development of modern theories of plastic deformation of glassy polymers.

Key Words: polymer glasses, relaxation, plastic deformation, structure

INTRODUCTION

Temperature-induced relaxation of plastically deformed polymer glasses has been a subject of many investigations [1 - 17]. In general case, temperature-induced relaxation involves recovery of residual deformation and relaxation of inner energy stored by polymer during its plastic deformation. This relaxation is realized in two well-pronounced temperature regions: at temperatures below polymer glass transition temperature T_g (i) and in temperature region of glass transition (ii). The existence of two relaxation modes allows one to suggest the existence of the corresponding modes of plastic deformation of glassy polymers. In connection with this, investigations focused on relaxation behavior of the deformed polymer glasses [1, 5, 7, 10, 17] provided a basis for the development of speculations concerning the mechanism of plastic deformation in glassy polymers.

Practical aspects of this problem seem to be quite evident. Deformation and relaxation behavior of polymer glasses are directly related to the development of advanced construction materials with improved properties, which are characetrized by combination of high rigidity and low brittleness. On the other hand, processing of polymer glasses into final articles is usually accompanied by polymer plastic deformation. Hence, shape stability and stability of properties of final articles appear to be intimately related to relaxation of this deformation in polymer under operating temperatures. In connection with this, the problems related to temperature-induced relaxation are of great practical importance for improvement in thermal stability of polymer materials based on polymer glasses [6].

At the present time, there is a vast of experimental evidence concerning temperature-induced recovery of residual deformation and relaxation of inner energy in the deformed polymer glasses. However, in many cases, experimental results obtained by different authors were in conflict with each other. This conflicting evidence suggests the necessity of a thorough examination of all available literature data concerning the experimental results as well as mechanisms of temperature-induced relaxation in deformed polymer glasses.

1. Experimental Evidence Concerning Temperature-Induced Relaxation in Deformed Polymer Glasses

Temperature-induced relaxation of residual deformation ε_{res} of plastically deformed polymer was studied for various polymer glasses [1 - 17]: initial and cross-linked PMMA [2 - 6], PS [7, 10], copolymers of styrene with methacrylic acid [1], PC [7, 10], amorphous PET [7, 10, 13], polysulfone [7], and cross-linked epoxy-amine polymers [7 - 16]. In [2 - 4], the test samples were uniaxially compressed at deformation temperatures T_{def} from 293 K to polymer glass transition temperature T_g. After that, they were cooled by liquid nitrogen in a stressed state. Temperature-induced recovery of ε_{res} was estimated by measuring the changes in height of the free-standing samples at heating with constant heating rate starting with 213K. In [7 - 16], the samples were deformed under uniaxial and hydrostatic compression and under creep condition at T_{def} by 70 - 250K below their T_g, and then the samples were unloaded at T_{def}. Temperature-induced relaxation of ε_{res} was also studied by measuring the changes in linear dimensions of the deformed samples at constant heating rate.

Despite a variety of polymers and regimes of plastic deformation, all regularities of temperature-induced relaxation of ε_{res} show general character.

In general case, recovery of geometrical sizes of plastically deformed polymer glasses takes place in two well-defined temperature regions and involves two components: low-temperature component ε_1 and high-temperature component ε_2 (Fig. 1). Let us note that high-temperature component of relaxation ε_2 is realized in rather narrow temperature region related to polymer glass transition, whereas low-temperature component ε_1 is associated with a wide temperature interval at temperatures below T_g. Minimal temperature, at which the appearance of low-temperature relaxation was revealed, is equal to 103 K [7]. Each temperature below T_g is associated with a certain recovery of ε_{res}, and the whole temperature dependence of ε_1 shows a well-pronounced smooth character (Fig. 1a).

The ratio between the low-temperature and high-temperature components of recovery of ε_{res} was studied as a function of strain [2, 5]. For the compressed samples with rather low strains complete recovery of ε_{res} takes place at temperatures below T_g (Fig. 2, curves 1,2). On the contrary, complete recovery of ε_{res} of the samples with high strains is observed on heating to T_g (Fig. 2, curves 3-5). However, in this case low-temperature component is also involved.

Figure 3 shows the dependences of low-temperature ε_1 and high-temperature ε_2 components on strain as well as the corresponding stress-strain diagram of uniaxial compression of PMMA. As follows from Figure 3, in the case of the samples with strains below yield point ε_y, recovery of ε_{res} proceeds via low-temperature mode ε_1. At the strain corresponding to ε_y, component ε_1 is maximal, and further increase in compression strain is associated with the appearance and growth in component ε_2 at constant ε_1.

As evidenced by DSC studies of deformed polymer glasses [1, 7, 10, 17, 18], on heating the samples of deformed polymer, a well-pronounced exothermic effect is observed, in contrast to initial undeformed sample (Fig. 4). The corresponding area of the exothermic peak in DSC curves provides a quantitative characterization of excess enthalpy ΔH_{exo} of the deformed polymer.

Examination of thermodynamic parameters of polymer deformation (work for deformation W, deformation heat Q_{def}, and variations in inner energy of the deformed polymer ΔU_{def}) in cyclic loading regime [8-10] showed that unloaded sample is characetrized by stored inner energy ΔU_{st}. The values of ΔH_{exo} estimated from the DSC curves for the deformed samples are quite similar to the values of ΔU_{st} as obtained by deformational calorimetry [9, 11] and measurements of heats of dissolution of deformed samples [11].

Hence, excess enthalpy may be related to inner energy stored in the deformed and unloaded samples. In connection with this, hereinafter, we will make no difference between these two parameters, and relaxation of enthalpy on heating of deformed samples will be treated as relaxation of inner energy.

Figure 5 shows the resolution of work for polymer deformation W into deformation heat Q_{def} and stored inner energy ΔU_{st} as estimated by deformational calorimetry [7, 10]. Let us emphasize that, at all stages of deformation studied, a marked fraction of W is stored in polymer as ΔU_{st}. This feature is characteristic of plastic deformation of glassy polymers. In the case of an ideal plasticity, work for deformation is completely transformed to Q_{def}. The behavior of most of plastic materials (primarily, crystalline metals) is close to the above pattern and changes in inner energy induced by plastic deformation are only few percent of W. For polymers, the corresponding value is estimated to be as high as tens percent.

Comparison between Figs. 3 and 5 suggests a formalized correlation of strain dependences of low-temperature component ε_1 of relaxation of residual deformation and stored inner energy ΔU_{st}. As was shown in [10, 12], almost all inner energy ΔU_{st} is related to low-temperature component ε_1 and relaxation of ΔU_{st} is observed at temperatures below T_g. However, this result appears to be in conflict with the speculations presented in [1, 18], according to which temperature-induced relaxation of stored inner energy shows a two-stage character, and complete relaxation of ΔU_{st} takes place at temperatures corresponding to temperature region of glass transition (Fig. 6, curve 1). In connection with this, there is no an evident correlation between the energy stored in polymer during its plastic deformation and low-temperature component ε_1 of relaxation of ε_{res}. Nevertheless, storage of inner energy in polymer is always associated with the appearance of residual deformation.

2. Modern Approaches to Description of Temperature-Induced Relaxation in Deformed Polymer Glasses

Hence, temperature-induced relaxation of plastically deformed polymer glasses involves two events. The first is related to recovery of residual deformation, which is associated with two well-defined temperature regions, whereas the second is provided by relaxation of stored inner energy ΔU_{st}, which is also characterized by a two-stage character [1]. To gain a deeper insight into the mechanism of temperature-induced relaxation of the deformed polymer glasses, let us discuss the nature of these phenomena and correlation between them. Let us start our discussion with possible reasons of changes in inner energy during deformation of glassy polymers.

In general case, storage of inner energy during polymer deformation may be related to the two following reasons [1, 19]:

1. Conformational (intramolecular) changes.

Plastic deformation of polymers, similar to rubbery deformation, is accompanied by a lowering in conformational entropy, and changes in intramolecular conformational energy. In contrast to rubbery deformation, plastic deformation of polymer glasses at temperatures below T_g is not associated with a complete recovery of energy intramolecular balance on polymer unloading. The reason is related to the frozen-in segmental mobility at temperatures below T_g.

For polymer chains with different conformational energy, i.e., with hindered inner rotation, the changes in intramolecular energy are dictated by transition from one conformational state to another ones. The tendency of changes in energy state of the deformed polymer (increase or decrease in inner energy) are controlled by the sign of temperature coefficient of unperturbed dimensions of macromolecule $d\ln\langle r^2\rangle_0/dT$, where $\langle r^2\rangle_0$ is a mean distance between the ends of free-standing polymer chain.

Positive values of $d\ln\langle r^2\rangle_0/dT$ are characteristic of polymers, in which more extended conformational sequences possess higher energy. As a result of plastic (or rubberlike elastic) deformation or with increasing temperature, the fractional content of such sequences increases, and inner energy of deformed polymer also increases.

In the case of polymers, in which extended conformations are characterized by lower conformational energies and $d\ln\langle r^2\rangle_0/dT < 0$, polymer deformation is associated with a decrease in inner energy.

Hence, in general case, polymer deformation may be accompanied either by increase or decrease in inner energy. In connection with this, transformation of energy conformational state during polymer deformation may hardly be considered as a universal factor responsible for an increase in inner energy of the deformed polymer.

2. Development of inner stresses.

In deformed polymers, a certain fraction of work for deformation is associated with the development of inner stresses in polymers. Relaxation of these inner stresses should be accompanied by scattering of the stored energy. The nature of strain-induced inner stresses will be discussed in more detail.

3. Development of new surfaces.

In general case, deformation of polymer bodies is accompanied by the development of a new surface as a result of microcracking and crazing as well as polymer fibrillation. However, this factor may be ignored taking into account small contributions from the above processes to the changes in energy state of the deformed polymer.

4. Disruption of chemical bonds.

Disruption of chemical bonds in polymers under their deformation is recognized by the appearance of free radicals in the material [20]. However, the number of disruption events is rather small, and the number of macromolecules involved does not exceed 10^{18} g^{-1}. The energy providing the formation of free radicals and energetics of concomitant mechanochemical processes are negligibly small, and this factor may be ignored.

5. Changes in volume of deformed polymer.

With increasing temperature from room temperature to glass transition temperature, direct measurements of changes in volume of PS samples after their uniaxial compression showed no marked changes in volume within an accuracy of $\Delta V/V = 10^{-4}$ [14, 21]. In this case, during heating, a marked release of the stored inner energy is observed. This evidence suggests that storage and release of inner energy for deformed glassy polymers may take place at small volume changes.

Hence, the above speculations concerning the possible sources providing changes in energy during polymer deformation allow us to conclude that the development of inner stresses and their storage in unloaded samples, subsequent relaxation of which is characetrized by exothermic character, is of primary importance.

According to the theory developed by Oleinik et al. [7, 10, 17], storage of inner energy in deformed glassy polymers is related to the fact that, under the action of external stresses, elementary local shear bands (plastic shear transformations) appear in polymer. Plastic shear transformations (PSTs) are nonconformational shear structures, each of which is characterized by local shear strain ε_{PST}^{loc} and is surrounded by field of elastic stresses. Local shear strains and field of elastic stresses corresponding to PSTs may exist only in polymer glasses because of their high viscosity and they are not allowed in rubbery state because of fast relaxation.

Accumulation of PSTs in glassy polymer proceeds at rather early stages of deformation, and as concentration of PSTs increases, inner energy of the deformed polymer also increases. At strains of about 25 - 30%, stationary stage of deformation of glassy polymer is realized, and this stage is characterized by constant high level of stored inner energy (Fig. 5, curve 4). This evidence is assumed to be controlled by equilibrium between nucleation and degradation of PSTs. Degradation of PSTs is associated with their transition into new chain conformations and severe conformational gauche-trans rearrangements. This process is accompanied by the appearance of extended coils enriched with nonequilibrium conformational sequences. Finally, local inner energy of PST disappears, but local strain remains preserved. Let us note that evolution of PSTs was invoked in [7, 10] to rationalize an increase in mechanical loss as well as an increase in enthalpies of activation of relaxation processes in deformed polymer glasses, as compared with initial samples.

Low-temperature relaxation of deformed polymer glasses at temperatures below T_g is provided by recovery of stored PSTs associated with their transition to initial state. In this case, both local inner energy and local shear strain of each PST disappear. This process is assumed [7] to have nothing in common with conformational changes in polymer chains. Speculations concerning nonconformational nature of low-temperature relaxation via recovery of PSTs are supported by direct spectral observations [13]. According to [13], relaxation processes in polymer glasses at temperatures below T_g do not involve conformational trans-gauche transitions[1]. High-temperature relaxation of residual deformation in the region of polymer glass transition is controlled by entropy recovery of nonequilibrium conformations of macromolecular coils via segmental mobility. Just within temperature region of glass transition, dramatic

[1] We would like to clarify this point. Within certain limits, conformational changes in polymer chain due to segmental mobility (transition from excited extended coil to equilibrium state) may be provided by torsional oscillations of groups of atoms around equilibrium state without any trans-gauche transitions within monomer unit [22]. In connection with this, we believe that the absence of trans-gauche transitions in temperature region associated with low-temperature relaxation does not provide an unequivocal evidence concerning nonconformational nature of this relaxation mode.

conformational rearrangements accompanied by the trans-gauche transitions were observed [13].

The problems concerning the scale of molecular motions responsible for development and relaxation of PSTs arise. According to [7], structural rearrangements in PSTs, which are able to relax at temperatures below polymer T_g, should be of small-scale order. As follows from experimental observations, the onset of low-temperature relaxation of ε_{res} was revealed at 103K. This allows one to conclude that, at such temperatures, structural rearrangements involved in low-temperature mode of relaxation can not be associated with large-scale (for example, segmental) motions in macromolecular chains [7][2].

In [15, 16], activation volumes corresponding to the development and relaxation of PSTs were estimated from sensitivity of these processes to shear stresses and external hydrostatic pressure. In the case of epoxide polymer, at temperatures from 293 to 363K, activation volume estimated from sensitivity of appearance and relaxation of PSTs to shear stress component was found to be equal to 0.25 - 0.4 nm^3. The corresponding values estimated from sensitivity of relaxation of PSTs to hydrostatic pressure component were 0.055 and 0.1 nm^3 for epoxy polymer and polycarbonate, respectively. The above values for activation volume of relaxation of PSTs correlate well with the corresponding values for the γ- and β-relaxation modes. This evidence allows one to conclude that low-temperature relaxation via recovery of PSTs is dictated by small-scale molecular motions. Let us note that activation

[2]We think that recovery of ε_{res} at such low temperatures does not provide a unequivocal evidence in favor of small-scale molecular motions responsible for low-temparture mode of relaxation. This problem will be discussed in more detail later.

volume for high-temperature component of recovery of residual deformation is close to that of the α-relaxation.

The speculations concerning the small-scale character of structural rearrangements associated with low-temperature component of temperature-induced relaxation of ε_{res} do not provide any explanation of a smooth character of temperature dependences of this mode of relaxation. Relaxation of PSTs should be sensitive to the appearance of γ- and β-relaxation modes on polymer heating, the activation volumes of which are comparable to the corresponding characteristics of relaxation of PSTs. In temperature region associated with the above relaxation modes, one should anticipate a sharp increase in recovery of ε_{res}. However, experimental curves show no marked contributions from any relaxation modes in recovery of ε_{res} in glassy state (Fig. 1a). The existence of a wide temperature region associated with relaxation of ΔU_{st} and smooth character of low-temperature component of recovery of ε_{res} is treated from the viewpoint of a wide distribution of activation energies of relaxation of each PST [7].

Let us also note that the theory of temperature-induced relaxation of the deformed polymer glasses based on speculations concerning PSTs has been developed for defect but uniform polymer medium with no account for specific structural features of glassy polymer.

During recent decades, the problems related to structure of amorphous glasses are being widely discussed in literature. Let us mention some of them.

As demonstrated by electron microscopy [23 - 30], electron diffraction [29, 31], positron annihilation [32], light scattering measurements [33], small-angle X-ray scattering [27], and wide-angle X-ray scattering [31], polymer glasses are characterized by structural inhomogeneity related to the existence of heterogeneous ordered regions with higher packing density (domains, clusters) with dimensions up to several nanometers. These domains are joined with each other by numerous tie chains, which constitute the regions with lower packing density and ordering. Evidently, an individual macromolecule appears to be involved in several domains and interdomain regions, and this factor is responsible for cooperative character of structure of glassy polymers.

Experimental evidence obtained provided a fundamental basis for the development of various structural models of glassy polymers. The difference between the models advanced is related to the type of principal morphological structural unit: fibril [4], domain [34, 35], globule [36 - 38], and cluster [39, 40]. However, at the present time, a certain lack of convincing experimental evidence does not allow one to decide in favor of this or that particular model. Hence, in this work, our speculations will be based only on unequivocal conclusion concerning structural inhomogeneity of glassy polymer, which is associated with the existence of local structural regions with different packing density and ordering of macromolecules or their fragments. This approach allows a plausible description of specific effect of plasticizing agents on molecular mobility in polymers [41. 42], dual-mode sorption of gases by glassy polymers [43 - 45], development of the irreversibly bound form of low-molecular-mass compounds in polymers [46 - 48]. The speculations concerning structural inhomogeneity are also invoked to rationalize mechanical properties of polymer glasses [23, 28, 43]. Let us mention some works [42, 49, 50, 51], in which specific dual transitions observed in temperature region of glass transition were treated in terms of structural regions with different packing density. In this case, low-temperature transition is attributed to the appearance of segmental mobility in loosely packed regions, whereas high-temperature transition is related to segmental mobility in densely packed domains. According to [51], for various polymers, the temperature regions associated with these transitions may be shifted by dozens of degrees.

To a certain extent, structural approach providing description of temperature-induced relaxation in the deformed polymer glasses was outlined in [1]. As a principal reason accounting for an increase in inner energy during deformation of glassy polymer, changes in energy of intermolecular interaction were considered. Comparing the data obtained by DSC and IR spectroscopy [18] allowed one to advance some ideas concerning the nature of changes in energetic state of glassy polymers under deformation. PS and copolymers of styrene with methacrylic acid (from 2 to 33 mol %) were studied. Let us note that the above copolymers are characetrized not only by universal Van-der-Waals intermolecular interaction by also by a specific mode of intermolecular interaction, which may be controlled by double and ordinary hydrogen bonds between carboxyl groups in methacrylic acid. The energies of these bonds are 38 and 19 kJ/mol, respectively.

In addition to calorimetric studies providing estimation of ΔU_{st}, IR spectroscopy was used to determine:

-parameter of universal intermolecular interaction related to interaction between benzene rings;

-energy parameter of specific intermolecular interaction associated with hydrogen bonds between carboxyl groups of methacrylic acid;

-parameter of intramolecular energy with respect to the ratio between absorption bands of gauche- and trans-isomers.

In the case of PS, an increase in inner energy during deformation was shown to be accompanied by a decrease in intensity of universal intermolecular interaction. Contribution from intramolecular energy to changes in polymer energy state was so small that ΔU_{st} was considered as measure of energy changes in intermolecular interaction.

In the case of deformation of copolymers of styrene with methacrylic acid, changes in universal intermolecular interaction was rather small, and an increase in inner energy during deformation was primarily associated with disruption of intermolecular hydrogen bonds [52 - 54]. As evidence by DSC scans, exothermic effect for deformed copolymers is controlled by recovery of hydrogen bonds with heating.

As follows from Fig. 6, the temperature dependences of relaxation of ΔU_{st} (curve 1) and recovery of hydrogen bonds (curve 2) show parallel profiles. Temperature dependence of recovery of linear dimensions of deformed samples (curve 3) demonstrates the similar behavior. All above processes proceed in two stages. The first stage is associated with the β-transition, whereas the second one is observed within temperature interval of polymer glass transition.

Hence, storage of inner energy in glassy polymers during their deformation is related to violation of intermolecular specific interactions [1]. (This conclusion correlated well with the speculations presented in [55]. According to [55], storage of inner stresses during the deformation of glassy polymer is primarily related to changes in intersegmental distances.) With increasing temperature, exothermic effect is attributed to the recovery of the disrupted intermolecular bonds. Let us note that low-temperature mode of relaxation of stored inner energy and recovery of residual deformation is directly related to the β-transition in polymer glasses [1]. To provide a more detailed description of the mechanism of relaxation of the deformed polymer glasses, let us consider correlation between the α- and β-transitions.

Speculations concerning the mechanism of the β-relaxation in glassy polymers and its correlation with the α-relaxation were advanced in [1] and based on DSC measurements on activation parameters of the β-transition in glassy polymers [56 - 58] reinforced with NMR data obtained for the loaded oriented polymer [59], long-wave IR spectroscopy [60 - 65] as well as with theoretical and spectral examination of the dynamics of hydrogen bonds [52, 53]. As was proposed, the principal structural kinetic unit responsible for both β- and α-transitions in glassy polymers is statistic (thermodynamic) segment. The mechanism of the β-transition involves quasi-independent and localized motion of statistic segment associated with overcome of intermolecular barriers and with participation of the trans-gauche transition in more loosely packed regions. The relaxation α-transition is accomplished via large-scale cooperative motions of the contacting neighboring segments in a whole volume of a polymer body. Hence, an event associated with the α-transition is nothing more but a cooperation of several events of the β-relaxation.

From this standpoint, the onset of structural relaxation of polymer glasses, which involves enthalpy relaxation and recovery of ε_{res}, corresponds to the temperature region associated with the β-relaxation (Fig. 6). Just at these temperatures, more loosely packed regions are involved in segmental mobility. A complete enthalpy relaxation and a complete recovery of ε_{res} of the deformed polymer glasses takes place at the temperature corresponding to the relaxation α-transition via segmental mobility in a whole polymer sample.

However, the data on recovery of ε_{res} in the deformed glassy polymers [1, 18] (Fig. 6, curve 3) appears to be in conflict with the corresponding data obtained in [2 - 4, 7]. As was mentioned earlier, recovery of ε_{res} of the deformed polymer glasses was observed starting with 213 K [2 - 4] and 103K [7]. This does not allows one to correlate the onset of recovery of ε_{res} with the onset of relaxation β-transition. Furthermore, the temperature dependence of ε_1 appears to be insensitive to the appearance of the β-relaxation (Fig. 1a). However, let us emphasize that, according to [1], low-temperature component ε_1 of recovery of residual deformation is attributed to the appearance of segmental mobility in structural regions with lower packing density.

In our opinion, the speculation concerning structural inhomogeneity in polymer glasses should have logical development. Two structural sublevels with higher and lower packing density as recognized by modern structural and physicochemical studies are limiting cases in structural organization of glassy polymer. We believe that there is a continuous spectrum of intermediate states (structural sublevels) with different packing density and ordering. Within this standpoint, the temperature dependence of recovery of ε_{res} for the deformed polymer glasses was examined [5, 6].

Each of these structural sublevels is associated with its own relaxation time and activation energy of segmental mobility. A continuous spectrum of packing density and local ordering within the sublevels of structural inhomogeneity of polymer glasses is responsible for a continuous spectrum of activation energies and relaxation times of segmental mobility in polymer glasses. Hence, at each temperature below T_g local segmental mobility within certain sublevels is allowed. As temperature increases, segmental mobility in structural sublevels with higher packing density comes into play, and, finally, at T_g all polymer sample appears to be involved in segmental motion.

In the case of the deformed polymer glasses, with increasing temperature, the appearance of segmental mobility in structural sublevels with higher packing density is accompanied by a gradual recovery of plastic deformation within this particular sublevels. A gradual appearance of segmental mobility in structural levels with higher packing density allows one to rationalize a smooth character of temperature dependence of ε_1 (Fig. 1a) as well as to explain the fact that each temperature is associated with a certain value of recovery of ε_{res}. A complete recovery of plastic deformation is accomplished at T_g, when the segments in densely packed structural regions (domains) are involved in segmental mobility (Fig. 1a, component ε_2). Hence, molecular mechanism for low-temperature relaxation is similar to that of high-temperature mode and involves transition of excited chain conformations to initial equilibrium state via segmental mobility in local structural sublevels in polymer glasses. Let us now consider the validity of this approach.

At the present time, the existence of structural sublevels with different packing densities in polymer glasses has been unequivocally proved by modern experimental methods [23 - 33], and the dimensions of free volume in loosely packed regions are comparable to activation volume of segments. This reasoning suggests a possible realization of segmental mobility in local structural sublevels at temperature below polymer glass transition point. (Direct observations indicating realization of segmental mobility in structural sublevels with lower packing density for polymer glassy state were reported in [1, 42, 49, 51]).

Additional factors assisting segmental mobility in local structural regions in glassy polymer are the following. First, asymmetry of the field of intermolecular interaction provided by free volume is associated with a decrease in activation energy of segmental mobility [66]. Second, a certain decrease in activation energy of segmental mobility may be also related to the existence of inner stresses in polymer stored during polymer processing or polymer deformation [2, 4]. Deformation of glassy polymer is accompanied by severe structural rearrangements and/or displacements in cooperative system of supermolecular (or supersegmental) structural elements. As a result, the development of inner (most likely, elastic) stresses is observed. Highly cooperative nature of structural organization in glassy polymers may be invoked to rationalize higher level of inner energy stored during polymer deformation, as compared with low-molecular-mass glassy and crystalline materials.

Hence, the advanced mechanism [5, 6] for temperature-induced relaxation in the deformed polymer glasses, which is based on speculation concerning a step-by-step appearance (with increasing temperature) of segmental mobility in structural sublevels with different packing density, seems to be quite plausible with respect to experimental evidence on structure, mechanical properties, and specific features of relaxation in polymer glasses. The only point, which lacks direct experimental support, is related to a continuous spectrum of packing density in polymer glasses. However, this assumption seems to be a further development of modern speculations concerning structural inhomogeneity in glassy polymers.

Finally, we would like to mention that the description of two-stage character of recovery of residual deformation via a gradual appearance of segmental mobility in structural regions with different packing density allows one to approach the problems of plastic deformation of polymer glasses from a somewhat different standpoint [5, 6]. At early stages (at $\varepsilon < \varepsilon_y$), plastic deformation is associated with a step-by-step mechanical activation of segments in polymer regions with lower packing density. This fraction of residual deformation is able to relax at temperatures below polymer glass transition temperature due to thermally activated segmental mobility within the above structural sublevels (low-temperature component ε_1). As strain increases, the segments within more densely packed polymer regions appear to be involved in mechanical activation. Relaxation of this fraction of ε_{res} is allowed only in the temperature region corresponding to glass transition (high-temperature component ε_2) and is realized via thermal activation of segments within the whole volume of polymer samples.

The above reasoning may serve as an extra argument in discussion concerning the interplay between diffusion (molecular kinetic) [67 - 69] and solid crystallike [70 - 72] theories for plastic deformation of polymer glasses. The first approach is related to mechanically activated liquidlike rearrangement in chain segments, which are similar to those in rubbers. Within the second approach, plastic deformation of polymer glasses is dictated by evolution of small-scale local shears, which are similar to defects in crystals. By analogy with deformation behavior of crystals (for example, [73, 74]), which is characterized by a well-pronounced transition from dislocation mechanism to diffusion mechanism at $T_{def} \geq 0.6 - 0.7\ T_m$, in [75], similar transition of polymer plastic deformation to diffusion mechanism was observed at temperatures above a certain temperature T_1. In [7, 10, 17], plastic deformation of glassy polymer at temperatures below T_g is considered in terms of the mechanism of a solid body, and the threshold temperature associated with the transition from solid-body mechanism to diffusion mechanism is provided by glass transition temperature. Speculations concerning structural inhomogeneity in polymer glasses, which accounts for a gradual involvement of polymer segments within local structural regions into stress-activated liquidlike rearrangements, allow one to advance a plausible description of plastic deformation in glassy polymers, which is based on universal molecular kinetic fundamentals without any appeal to crystallike theories concerning deformation of solid bodies.

CONCLUSION

Temperature-induced relaxation in plastically deformed polymer glasses is accompanied by recovery of residual deformation in two well-pronounced temperature intervals at temperatures below T_g and in the temperature interval of polymer glass transition as well as by relaxation of the stored inner energy. At the present time, the mechanism of temperature-induced relaxation of the deformed polymer glasses is treated in terms of two alternative approaches.

The first approach [7, 10, 17] is based on speculations concerning independent recovery of elementary local shear bands (plastic shear transformations) at temperatures below polymer glass transition temperature and entropy recovery of nonequilibrium conformations of macromolecular coils via realization of segmental mobility in polymer at temperatures corresponding to polymer glass transition.

An alternative approach involves the speculations concerning structural inhomogeneity in polymer glasses, that is, the existence of local structural regions with different packing density and ordering [1, 2, 4 - 6]. According to this approach, the recovery of residual deformation at temperatures below T_g is rationalized in terms of a gradual involvement of structural regions with lower packing density into segmental mobility [5, 6]. A complete relaxation of plastically deformed polymer glasses is accomplished at temperatures associated with polymer glass transition via realization of segmental mobility in a whole polymer sample.

Even though at the present time there is no one universal theory accounting for temperature-induced relaxation in the deformed polymer glasses, let us emphasize a indisputable importance of this direction of polymer physics. A sophisticated pattern of relaxation in plastically deformed glassy polymers is a macroscopic manifestation of a set of physical processes, which are responsible for and/or accompany plastic deformation of this class of polymer materials. In this work, we discussed some approaches to the description of deformation and relaxation behavior of polymer glasses in terms of solid-body [7, 10, 17] and diffusion theories [1, 5, 6]. In our opinion, these approaches are not conflicting but complementary with respect to each other. Evidently, in the case of polymer glasses, which take intermediate position between liquids and true solid bodies, one should reasonable anticipate a superposition of solid-body and liquidlike diffusion mechanisms of plastic deformation and relaxation. This intriguing problem is still unclear and generates much interest from the viewpoint of the development of a mechanics of this class of polymers.

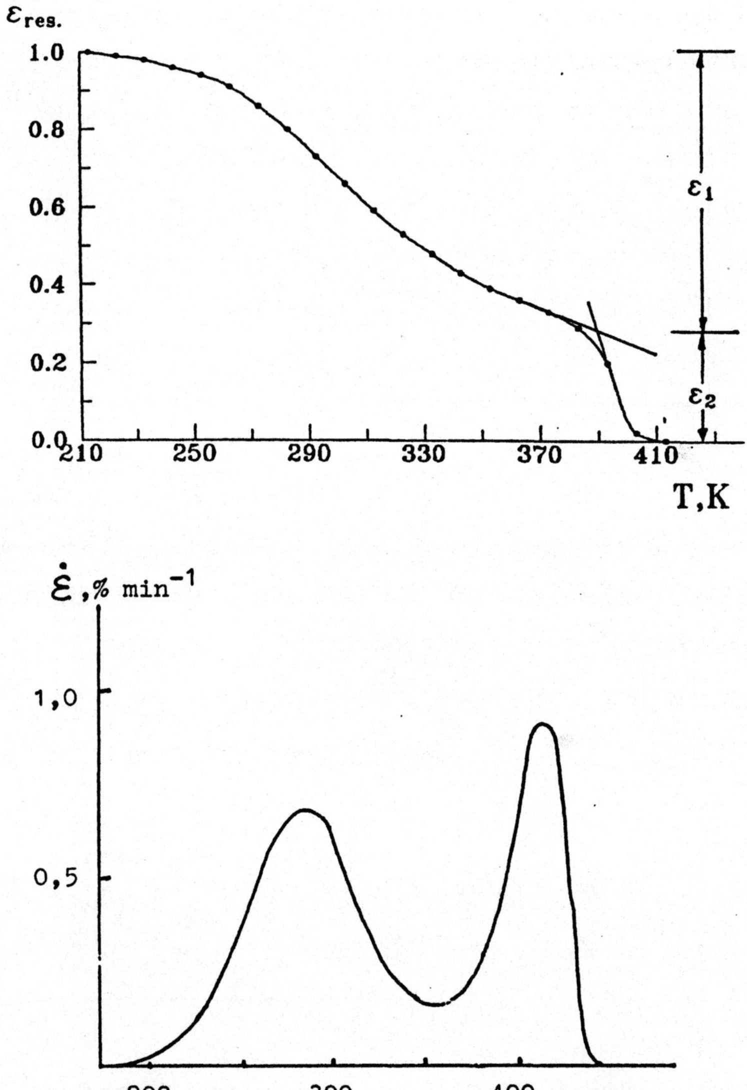

Fig. 1. Temperature dependences of relative residual deformation ε_{res} of PMMA (T_{def} = 293K, ε = 20%, uniaxial compression) [2] (A) and rate of recovery of residual deformation ε_{res} of epoxy-amine polymer (T_{def} = 188K, ε = 25%, uniaxial compression under creep conditions) [10] (B).

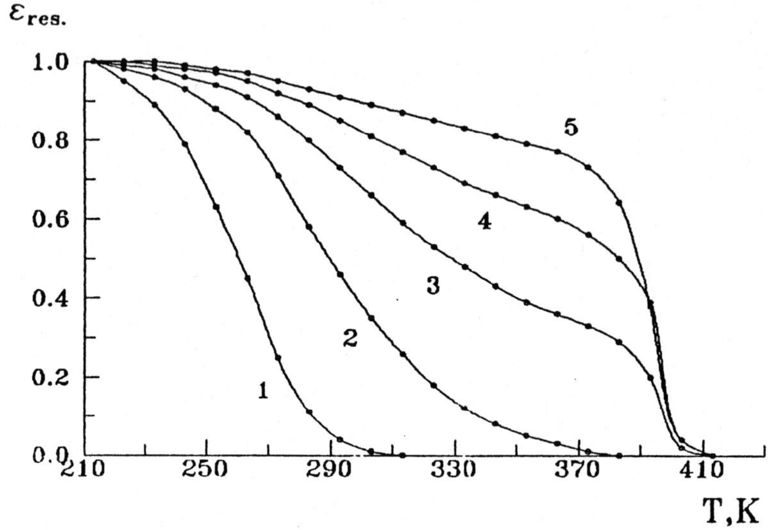

Fig. 2. Recovery of residual deformation ε_{res} of PMMA samples with strains 10 (1), 14 (2), 20 (3), 30 (4), and 50% (5). (T_{def} = 293K, uniaxial compression) [2, 5].

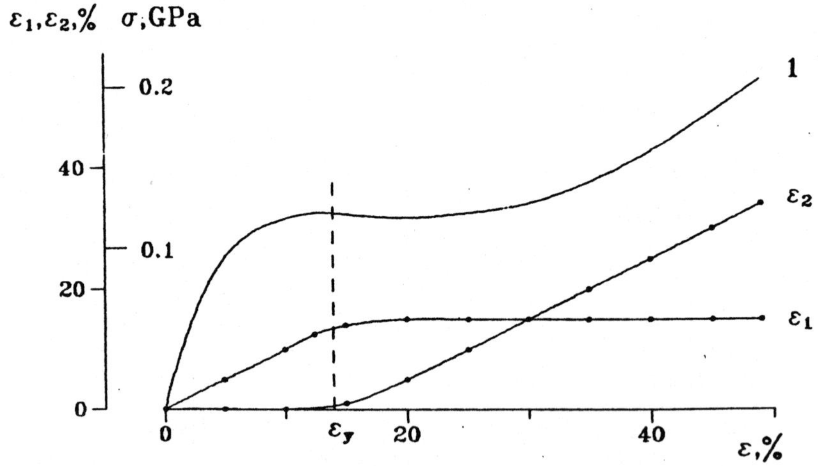

Fig. 3. Stress-strain curve corresponding to uniaxial compression of PMMA samples at 293K (curve 1) and low-temperature ε_1 and high-temperature ε_2 components of temperature-induced recovery of residual deformation versus strain ε [2, 5].

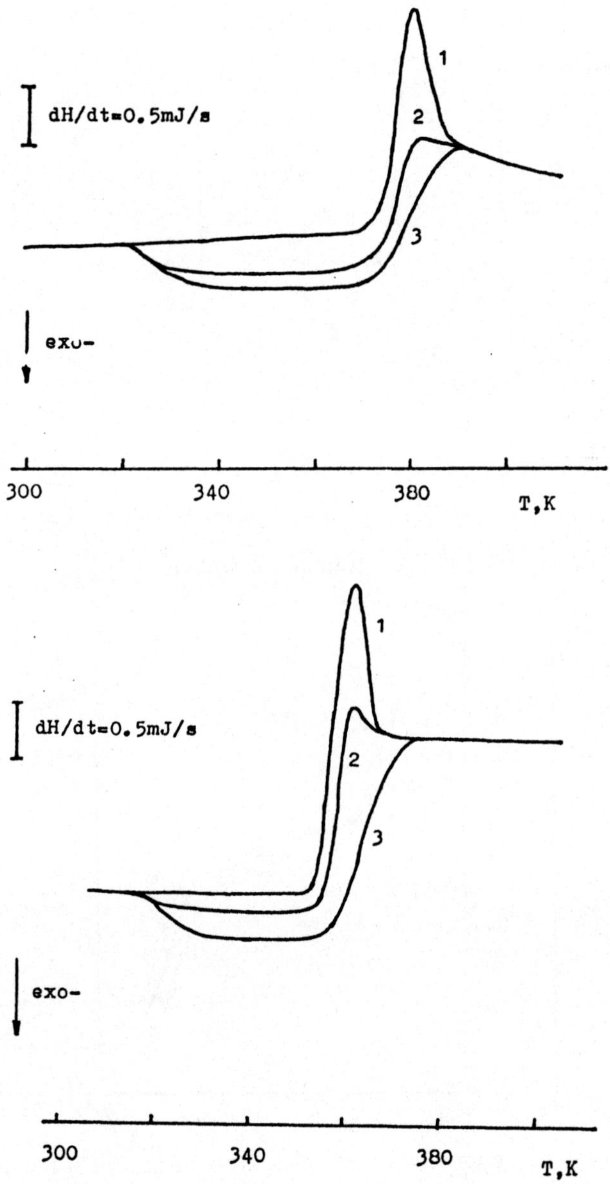

Fig. 4. DSC curves of virgin PS (1) and PS uniaxially compressed to strains 20 (2) and 40% (3) (A); and virgin PVC (1) and PVC uniaxially drawn (2) and compressed (3) to strain 40% (B) [1].

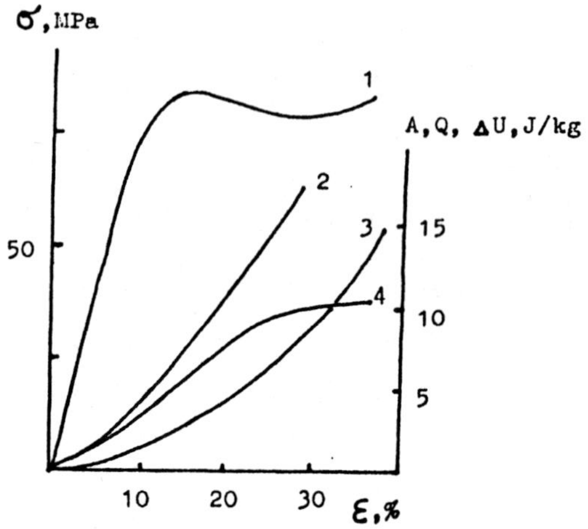

Fig. 5. Stress-strain curve corresponding to uniaxial compression of PS at 303K (1) and mechanical work for deformation W (2), heat of deformation Q (3), and inner energy ΔU (4) as a function of strain ε [10].

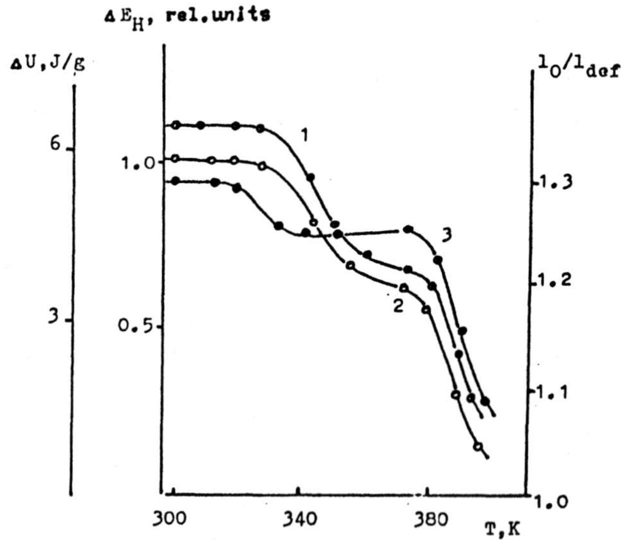

Fig. 6. Temperature dependences of relaxation of stored inner energy ΔU_{st} (1), recovery of hydrogen bonds ΔE_H (2), and recovery of linear dimensions l_0/l_{def} (3) of copolymer of styrene with 16 mol.% of methacrylic acid (T_{def} = 293K, ε = 30%, uniaxial compression) [1, 18].

REFERENCES

1. V.A. Bershtein and V.M. Egorov, Differentsial'naya Skaniruyushchaya Kalorimetriya v Fizikokhimii Polimerov (Differential Scanning Calorimetry in Physical Chemistry of Polymers), Leningrad: Khimiya, 1990 (in Russian).

2. S.A. Arzhakov, Doctorate (Chem.) Dissertation, Moscow: Karpov Institute of Physical Chemistry, 1975 (in Russian).

3. S.A. Arzhakov and V.A. Kabanov, Vysokomol. Soedin., B, 13, 318 (1971) (in Russian).

4. S.A. Arzhakov, N.F. Bakeev and V.A. Kabanov, Vysokomol. Soedin., A, 15, 1154 (1973) (in Russian).

5. M.S. Arzhakov and S.A. Arzhakov, Intern. J. Polymeric Mater., (1995) (in press).

6. M.S. Arzhakov, S.A. Arzhakov and V.A. Chernavin, New Polymeric Mater., (1996) (in press).

7. E.F. Oleinik, O.B. Salamatina, S.N. Rudnev and S.V. Shenogin, Polymer Science, 35, 1532 (1993).

8. S.N. Rudnev, O.B. Salamatina, V.V. Voenniy and E.F. Oleinik, Colloid Polym. Sci., 269, 460 (1991).

9. O.B. Salamatina, S.N. Rudnev and E.F. Oleinik, J. Therm. Anal., 38, 1271 (1992).

10. E.F. Oleinik, Progr. Colloid Polym. Sci., 80, 140 (1989).

11. O.B. Salamatina, S.I. Nazarenko, S.N. Rudnev and E.F. Oleinik, Mekhanika Kompozitnykh Materialov, 979 (1988) (in Russian).

12. E.F. Oleinik, High Performance Polymers, Eds., E. Baer and S. Moet, Minich: Hauser, 1990.

13. S.S. Sheiko, O.B. Salamatina, S.N. Rudnev and E.F. Oleinik, Vysokomol. Soedin., A, 32, 1552 (1990) (in Russian).

14. A.A. Gusev, S.I. Beshenko and E.F. Oleinik, Dokl. Akad. Nauk SSSR, 288, 1143 (1986) (in Russian).

15. I.Sh. Magafurov, V.A. Topolkaraev, R.E. Markaryan, A.L. Kovarskii and E.F. Oleinik, Vysokomol. Soedin, B, 32, 147 (1991) (in Russian).

16. I.Sh. Magafurov, Cand. Sci. (Phys.-Math.) Dissertation, Moscow: Semenov Institute of Chemical Physics, 1991 (in Russian).

17. E.F. Oleinik, O.B. Salamatina, S.N. Rudnev and S.V. Shenogin, Polymers for Advanced Technologies, 6, 1 (1995).

18. V.A. Bershtein, V.M. Egorov, L.G. Razgulyaeva and V.A. Stepanov, Vysokomol. Soedin., A, 20, 2278 (1978) (in Russian).

19. Yu.K. Godovskii, Teplofizika Polimerov (Thermal Physics of Polymers), Moscow: Khimiya, 1982 (in Russian).

20. H.H. Kausch, Polymer Fracture, Berlin, Heidelberd, New York: Springer, 1978.

21. E. Oleinik, Polym. J., 19, 105 (1987).

22. S.E. Bresler and Ya.I. Frenkel', Zh. Eksp. Teor. Fiz., 9, 1094 (1939) (in Russian).

23. K. Neki and P.H. Geil, J. Macromol. Sci., B, 8, 295 (1973).

24. H.E. Bair, M. Matsuo, W.A. Salmon and T.K. Kwei, Macromolecules, 5, 114 (1972).

25. J.L. Racich and J.A. Koutsky, J. Appl. Polym. Sci., 20, 2111 (1976).

26. M. Atsuta and D.T. Turner, J. Mater. Sci., Lett., 1, 167 (1982).

27. G.S.J. Yeh, IUPAC Macro 26th Symp., Mainz, Prepr. Short Commun., 2, 1176 (1979).

28. F.N. Kelley and D.R. Trainor, Polym. Bull., 7, 369 (1982).

29. P.H. Geil, Proc. 17th Intern. Symp., New York, London, 83 (1987).

30. D.A. Topchiev, Vysokomol. Soedin., A, 32, 2243 (1990) (in Russian).

31. J. Wendorff, Polymer, 23, 543 (1982).

32. V.P. Shantarovich, M.S. Arzhakov and I.B. Kevdina, Polymer Science, B, 37, 171 (1995).

33. N. Tanio, Y. Koike and Y. Ohtsuka, Polym. J., 21, 259 (1989).

34. G.S.J. Geil, J. Macromol. Sci., B, 6, 465 (1972).

35. I.J. Klement and P.H. Geil, J. Macromol. Sci., B, 5, 505 (1971).

36. Yu.A. Mikheev and L.N. Guseva, Khim. Fiz., 10, 724 (1991) (in Russian).

37. Yu.A. Mikheev, Intern. J. Polymeric Mater., 16, 221 (1992).

38. Yu.I. Matveev and A.A. Askadskii, Vysokomol. Soedin., A, 28, 1365 (1986) (in Russian).

39. G.V. Kozlov, D.S. Sanditov and V.D. Serdyuk, Polymer Science, A, 35, 1742 (1993).

40. V.N. Belousov, G.V. Kozlov, A.K. Mikitaev and Yu.S. Lipatov, Dokl. Akad. Nauk SSSR, 313, 630 (1990) (in Russian).

41. I.I. Perepechko and A.V. Prokazov, Mekhanika Polimerov, 749 (1977) (in Russian).

42. V.I. Surovtsev, S.S. Pelishenko, N.M. Zelenskaya-Surovtseva, A.A. Savitskaya and V.A. Degtyarev, Vysokomol. Soedin., A, 31, 339 (1989) (in Russian).

43. W.J. Koros and D.R. Paul, J. Polym. Sci., B, 16, 1947 (1978).

44. S. Subramanian, J.C. Heydweiller and S.A. Stern, J. Polym. Sci., B, 27, 1209 (1989).

45. W.R. Veith, J.M. Howell and H. Hsieh, J. Membr. Sci., 1, 177 (1976).

46. A.Ya. Polishchuk, L.A. Zimina, N.N. Madyuskin and G.E. Zaikov, Polymer Science, A, 35, 80 (1993).

47. M. Lewin, H. Guttman and Y. Naor, J. Macromol. Sci., A, 25, 1367 (1988).

48. M.S. Arzhakov, Intern. J. Polymeric Mater., 28, 103 (1995).

49. V.N. Belousov, B.Kh. Kotsev and A.K. Mikitaev, Dokl. Akad. Nauk SSSR, 280, 1140 (1985) (in Russian).

50. I.I. Perepechko and O.V. Startsev, Akusticheskii Zhurnal, 22, 749 (1976) (in Russian).

51. I.I. Perepechko, Akusticheskie Metody issledovaniya Polimerov (Acoustic Studies of Polymers), Moscow: Khimiya, 1973 (in Rissian).

52. V.A. Bershtein and N.A. Pertsev, Acta Polym., 9, 575 (1984).

53. N.A. Pertsev and V.A. Bershtein, Vysokomol. Soedin., A, 24, 1866 (1982) (in Russian).

54. V.A. Bershtein, L.G. Razgulyaeva, A.B. Sinani, and V.A. Stepanov, Fiz. Tverd. Tela, 18, 3017 (1976) (in Russian).

55. M. Nagasawa, A. Koizuka, K. Matssura, and M. Horita, Macromolecules, 23, 5079 (1990).

56. V.A. Bershtein, V.M. Egorov, and V.A. Stepanov, Dokl. Akad. Nauk SSSR, 269, 627 (1983) (in Russian).

57. V.A. Bershtein, V.M. Egorov, Yu.A. Emelyanov, and V.A. Stepanov, Polym. Bull., 29, 98 (1983).

58. V.A. Bershtein and V.M. Egorov, Vysokomol. Soedin., A, 27, 2440 (1985) (in Russian).

59. E.A. Egorov and V.V. Zhizhenkov, J. Polym. Sci., B, 20, 1089 (1982).

60. V.A. Bershtein and V.A. Ryzhov, J. Macromol. Sci., B, 23, 271 (1084).

61. V.A. Bershtein, V.M. Egorov, and V.A. Ryzhov, Vysokomol. Soedin., B, 28, 268 (1986) (in Russian).

62. V.A. Bershtein and V.A., Ryzhov, Dokl. Akad. Nauk SSSR, 284, 890 (1985) (in Russian).

63. V.A. Ryzhov and V.A. Bershtein, Vysokomol. Soedin., A, 29, 1852 (1987) (in Russian).

64. V.A. Ryzhov and V.A. Bershtein, Vysokomol. Soedin., A, 31, 451 (1989) (in Russian).

65. V.A. Ryzhov and V.A. Bershtein, Vysokomol. Soedin., A, 31, 458 (1989) (in Russian).

66. G.V. Kozlov, V.N. Shogenov, and A.K. Mikitaev, Dokl. Akad. Nauk SSSR, 298, 142 (1988) (in Russian).

67. V.A. Kargin, Izbrannye Trudy (Selected Works), Moscow: Nauka, 1986 (in Russian).

68. Yu.S. Lazurkin and R.L. Fogel'son, Zh. Tekhn. Fiz., 21 267 (1951) (in Russian).

69. R.E. Robertson, J. Chem. Phys., 44, 3950 (1966).

70. A.S. Argon, Phil. Mag., 28, 839 (1973).

71. P. Bowden and S. Raha, Phil. Mag., 29, 149 (1974).

72. B.A. Escaig, Polym. Eng. Sci., 24, 737 (1984).

73. A.S. Argon, Phys. Chem. of Solids, 43, 945 (1962).

74. V.I. Vladimirov, Fizicheskaya Teoriya Plastichnosti i Prochnosti (Physical Theory of Plasticity and Strength), Leningrad: Lenpolitekh, 1975 (in Russian).

75. J. Haussy, J.P. Cavrot, B. Escaig, and J.M. Lefebvre, J. Polym. Sci., B, 18, 311 (1980).

Studies in Deformed Poly(Methyl Methacrylate) Using Positron Annihilation Lifetime Spectroscopy

I.B. Kevdina, M.S. Arzhakov and V.P. Shantarovich

*Semenov Institute of Chemical Physics, Russian Academy of Sciences
ul. Kosygina 4, Moscow, 117977 Russia
**Polymer Department, Faculty of Chemistry, Moscow State University
Vorob'evy Gory, Moscow, 119899 Russia

Structural rearrangements in glassy PMMA on polymer plastic deformation and recovery of residual plastic deformation in glassy state were studied by positron annihilation lifetime spectroscopy. Uniaxial compression was shown to be accompanied by a decrease in concentration of free volume microregions in disordered polymer regions, which were characterized by low packing density. Recovery of residual deformation at elevated temperatures but below glass transition point T_g proceeds without any noticeable changes in fractional content of free volume both in disordered and ordered polymer regions. The advantages of positron annihilation lifetime spectroscopy for studying microstructure and structural rearrangements in polymers were discussed.

Key Words: PMMA, positron annihilation, deformation, relaxation, structure

INTRODUCTION

In recent years, many publications concerning studying elementary free volumes in polymer materials by positron annihilation lifetime spectroscopy appeared [1 - 6]. The advantages of this method are related not only to estimation of a certain correlation between annihilation characteristics and intensity of external field but also to providing a quantitative description of polymer microstructure. This approach necessitates development of the proper mathematical apparatus and specific physical models such as the model of entrapment of positron and positronium by defects proposed by Brandt even in 70s [7, 8], speculations concerning preferential localization of positrons in ordered polymer regions and positronium (electron - positron system) in disordered regions, the corresponding kinetic equations [1], finding two positronium components in lifetime spectra of positron annihilation (lifetime of one of the components is higher than 10 nsec), and calculations of mean radius of defect from lifetime characteristics [9, 10]. Computer program CONTIN was advanced to obtain a continuous distribution of defects in polymer material from positron annihilation lifetime spectra [4].

In this work, positron annihilation lifetime technique was used to study evolution of elementary free volumes during plastic deformation of glassy PMMA and subsequent strain recovery. Note that this method has been already used to study temperature transitions in PMMA in a temperature region from 73 to 523 K as well as polymer microstructure [6].

EXPERIMENTAL

We studied commercial PMMA prepared by bulk polymerization

(M_η=1.5x10^6). Before testing the cylindrical samples of PMMA (height 5mm, diameter 5 mm) were annealed at glass transtion temperature (393 K) for 1 h and then cooled to room temperature (293 K) with a cooling rate of 2 K/min.

Polymer deformation was carried out by uniaxial compression using an UTS-10 testing machine (Germany) at 293 K with a strain rate of 0.1 mm/min (2%/min) to a strain of 30%. Then, the samples were unloaded with the same rate. Residual plastic deformation ε_{res} was estimated as $\varepsilon_{res} = (h_o - h_{def})/h_o$, where h_o - initial height of the sample, h_{def} - the height of deformed sample.

After unloading, polymer samples were allowed to relax in a free state at 333 and 363 K. Recovery of residual deformation was studied by measuring the changes in height of the test samples until a constant value of $h_{rel\infty}$. Residual plastic deformation after relaxation was estimated as $\varepsilon_{res}^{rel}=(h_o - h_{rel\infty})/h_o$. The results are summarized in Table 1.

Lifetime spectra of positron annihilation (lifetime and intensity) were obtained at 293 K using an ORTEC spectrometer with a resolution function of 225 ps. Contribution to positron annihilation from $^{22}NaCl$ source in nickel foil did not exceed 20% and was accounted for by calculating program. Annihilation characteristics τ_i and I_i as calculated by computer program POSITRONFIT [11] are presented in Table 2.

Variations in fractional content and dimensions of free volume microregions induced by compression and following relaxation of PMMA samples 333 and 363 K were calculated according to a procedure described in [1, 6, 9]. In this case, local free volumes with positron (component with lifetime τ_2 and intensity I_2) and positronium (component τ_3 and I_3 as well as components τ_4 and I_4 in the case of four-component resolution) were presented as a three-dimensional potential well with a finite depth, and entrapment by defect was described by the corresponding kinetic equations assuming preferential localization of positrons in the polymer regions with increased packing density and of positronium in the regions with reduced packing density.

Table 3 shows the data on the rate of free positron annihilation λ_f, concentration N_+, and effective dimensions R_+ of free volume regions in ordered structural regions of PMMA, which are characterized by higher packing density as well as the corresponding values of N_{Ps} and R_{Ps} for disordered structural regions with lower packing density.

RESULTS AND DISCUSSION

As was mentioned earlier, the number of defects N_+ and N_{Ps} is approximated by the corresponding intensities of spectral components I_2 and I_3, and effective radii R_+ and R_{Ps} are characterized by lifetimes τ_2 and τ_3. Quantitative relationships of these parameters [1] also involve the rate of free positron annihilation and the fraction of positrons involved in a positronium for a given polymer Q. Mobility of positron and positronium determined by the corresponding diffusion coefficients D_+ and D_{Ps} is also taken into account. In the case of PMMA, they are equal to 0.1 - 1 and 10^{-4} cm^2/s, respectively [1. 7, 8]. When modern equipment with high resolution is used, an exact determination of Q presents substantial difficulties because of a variety of annihilation processes involved. When spectral resolution of singlet positronium and annihilation of non-localized positrons is not allowed, and annihilation of triplet positronium before its localization is not prevented (they contribute to the shortest-lived component I_1), then an exact determination of Q is impossible. However, for PMMA samples, $Q > 4I_3/3$, and in the case of initial sample, this is equal to 32%. Evidently, no changes in fraction of positrons involved in positronium are induced by compression, but formation of positronium is prevented by elimination of free microvolumes.Hence, Q is assumed to be equal to 40%, and this value is similar to that used in [12, 13] for PE.

Table 3 shows that on polymer deformation slight changes in N_+ (approximately from $0.7x10^{16}$ to $1.0x10^{16}$ cm^{-3}) lie within anticipated experimental error (20%). At the same time, a two-fold decrease in N_{Ps} is observed: from $1.5x10^{19}$ to $0.7x10^{19}$ cm^{-3}. This evidence allows one to suggest that deformation of PMMA (transition from sample A to sample B) is accompanied by decrease in fractional free volume in disordered polymer regions, and the net decrease in free volume is accomplished via decrease in fractional content (concentration) of free volume regions (N_{Ps}, Table 3) not via changes in dimensions of these regions (R_{Ps}, Table 3).

Recovery of residual plastic deformation in PMMA at elevated temperatures in glassy state (transition from sample B to samples C and D) proceeds without any noticeable changes in concentration and dimensions of free volume microregions both in ordered (N_+, R_+) and disordered (N_{Ps}, R_{Ps}) polymer regions (Table 3).

However, the dimensions of free volume microregions in loosely packed PMMA regions, as estimated from effective radius R_{Ps}, are equal to 0.3 - 0.4 nm^3 for all samples studied. By order of magnitude, this value coincides with activation volume of relaxation segmental transition (0.8 - 1.0 nm^3) and with the volume of statistic segment of PMMA (0.9 nm^3) [14]. This fact may serve as a supporting evidence that recovery of residual plastic deformation in PMMA at $T < T_g$ proceeds via segmental motion in disordered polymer regions.

Hence, positron annihilation lifetime spectroscopy provides a reliable means for studying polymer microstructure as well as structural transformation in polymer, in particular in PMMA, induced by deformation.

Table 1.

Deformation Characteristics of PMMA Samples.

Sample	h_0,mm	h_{def},mm	ε_{res},%	T_{rel},K	h_{rel},mm	ε_{res}^{rel},%
A	Initial polymer sample with a given prehistory					
B	5.0	4.2	16.5			
C	5.2	4.3	17.3	333	4.6	11.5
D	4.9	4.0	18.4	363	4.4	10.2

Table 2.

Characteristics of positron annihilation lifetime spectra of PMMA samples

Sample	τ_1,ns	I_1,%	τ_2,ns	I_2,%	τ_3,ns	I_3,%
A	0.242	45.0	0.514	32.2	1.93	22.8
	±0.012	±5.0	±0.039	±4.7	±0.03	±0.7
B	0.205	50.3	0.416	32.0	1.86	17.7
	±0.015	±8.1	±0.043	±7.7	±0.04	±0.6
C	0.227	48.0	0.457	31.7	1.85	20.3
	±0.015	±7.0	±0.045	±6.9	±0.04	±0.7
D	0.215	53.0	0.469	28.9	1.91	18.1
	±0.011	±6.0	±0.043	±5.1	±0.04	±0.6

Table 3.

Structural parameters of PMMA samples calculated from positron annihilation lifetime spectra

Sample	λ_f,s^{-1}	R_+,10^{-8}cm	N_+,$10^{16}$$cm^{-3}$	R_{Ps},10^{-8}cm	N_{Ps},$10^{19}$$cm^{-3}$
A	2.46	3.4	0.7	4.1	1.46
B	3.11	3.4	1.0	4.5	0.66
C	2.67	3.4	0.6	4.2	0.76
D	2.97	3.6	0.9	4.4	0.67

REFERENCES

1. A.V. Goldanskii, V.A. Onischuk and V.P. Shantarovich, Phys. Status Solidi, A, 102, 559 (1987).

2. T. Govorek, Phys. Status Solidi, A, 102, 511 (1987).

3. Yu.P. Yampolskii, V.P. Shantarovich, F.P. Chernyakovskii, A.I. Kornilov and N.A. Plate, J. Appl. Polym. Sci., 47, 85 (1993).

4. Q. Deng and Y.C. Jean, Macromolecules, 26, 30 (1993).

5. P. Kindl and P. Reiter, Positron Annihilation, Singapore: World Scientific, 1989.

6. I.B. Kevdina, Yu.M. Sivergin and V.P. Shantarovich, Khim. Vys. Energ., (1996) (in press).

7. W. Brandt and M. Mourino, Bull. Am. Phys. Soc., 20, 72 (1979).

8. W. Brandt and R. Paulin, Phys. Rev., 5, 2430 (1972).

9. V.P. Shantarovich, Yu.P. Yampolskii and I.B. Kevdina, Khim. Vys. Energ., 53 (1994) (inRussian).

10. Y. Kobayashi, K. Haraya, Y. Kamiya and S. Nattory, Bull. Chem. Soc. Jpn., 65, 160 (1992).

11. P. Kirkegaard and M. Eldrup, Comp. Phys. Commun., 3, 240 (1972); 7, 401 (1974).

12. F.J. Balta Calleja, J. Serna, J. Vincente and M.A. Segovia, J. Appl. Phys., 58, 253 (1985).

13. A.V. Gol'danskii, V.A. Onishchuk, V.P. Shantarovich, M.S. Akutin, N.B. Enikolopova and E.D. Lebedeva, Dokl. Akad. Nauk SSSR, 298, 633 (1988).

14. V.A. Bernshtein and V.M. Egorov, Differentsial'naya Skaniruyushchaya Kalorimetriya v Fizikokhimii Polimerov (Differential Scanning Calorimetry in Physical Chemistry of Polymers), Leningrad: Khimiya, 1990.

Structural and Mechanical Behavior of Polymethylmethacrylate containing Diethyl Siloxane Oligomer

M.S. Arzhakov, S.A. Arzhakov, V.V. Gustov,
I.B. Kevdina, and V.P. Shantarovich

*Polymer Department, Faculty of Chemistry, Lomonosov Moscow State University

Vorob'evy Gory, Moscow, 119899 Russia

**Semenov Institute of Chemical Physics,

Kosygina, 4, Moscow, 117977 Russia

Using the method of positron annihilation, structural changes induced by introduction of diethyl siloxane oligomers to PMMA were studied. The effect of these structural changes on the mechanical behavior of PMMA was discussed in terms of structural inhomogeneity of PMMA: existence of structural sublevels with different packing density and ordering. Elastic modulus, yield stress, and contribution of low-temperature component to the temperature-induced relaxation of plastic deformation were shown to be controlled by segmental mobility in structural sublevels with a lower packing density.

KEY WORDS polymer glasses, PMMA, low-molecular-mass additives, structure, mechanical properties

INTRODUCTION

Structural pattern of plastic deformation and furthertemperature-induced relaxation of glassy polymers was discussed in [1, 2]. Structure of glassy polymer is presented as a set of structural sublevels with different packing density and ordering. Local segmental mobility within a certain structural sublevel is controlled by local packing density in this region. From this standpoint, mechanical behavior of glassy polymers may be treated in terms of gradual involvement of various structural sublevels to deformation and relaxation. In this case, any changes in initial polymer structure and concomitant changes in packing density and ordering within the above sublevels should have a certain effect on both mechanical and relaxation properties of glassy polymers.

Structural changes in polymer may be accomplished by introduction of incompatible low-molecular-mass compounds, which are unable to produce a uniform distribution in polymer volume. This approach was formalized in terms of structural plasticization of polymers [3, 4]. Structural plasticization may be rationalized as follows: a low-molecular-mass compound with a low thermodynamic affinity to polymer is localized within local structural regions at boundaries between supramolecular or suprasegmental structures of glassy polymer. In this case, within these regions, changes in mobility of macromolecules or their fragments is observed, and this has a certain effect on mechanical behavior of polymer material. For PMMA, diethyl siloxane oligomers (DES) can be used as such low-molecular-mass modifying agents [5].

In this work, structural changes in PMMA induced by introduction of DES were studied, and the effect of these structural changes on polymer mechanical response was examined. The aim of this work was to elucidate the role of different levels of structural inhomogeneity of glassy polymer in polymer mechanical behavior.

As was shown in recent works [6 - 8], the method of positron annihilation is widely used to study polymer structure. Model experiments showed that positrons and positronium occupy ordered and disordered polymer regions, respectively. Maximal dimensions of structural inhomogeneities in polymer detected by this method are equal to tens of nanometers and are controlled by mobility of positron and positronium in ordered regions. Taking into account that structure of most of glassy polymers satisfy this requirement [the dimensions of ordered regions do not exceed several nanometers (for example, [9, 10])], concentration and effective dimensions of free volume microregions in ordered and disordered sublevels in oligo(ester acrylate) as well as the effect of intermolecular interaction and size of oligomer blocks on the content of free volume microregions were examined [8]. This experimental evidence was used to study structural changes in PMMA during plastic deformation and temperature-induced relaxation [7]. The method of positron annihilation was also used for characterization of structural changes in polymers containing low concentration of low-molecular-mass additives [11, 12].

EXPERIMENTAL

We studied PMMA prepared by bulk polymerization (M_W =1,500,000). Structural modification of PMMA samples involved introduction of diethyl siloxane oligomer (DES) with 5 repeat units during polymerization. The content of DES in PMMA was 0.1, 0.5, and 1.3 wt %.

Glass transition temperature was estimated by differential scanning calorimetry using a DuPont 1090 thermal analyzer. Heating rate was 20 K/min. Glass transition temperature was equal to 393 K for all samples.

Constancy of glass transition temperature at all concentrations of DES in PMMA as well as turbidity of all PMMA samples, containing DES, suggest a thermodynamic incompatibility between this low-molecular-mass compound and polymer matrix.

Thermostimulated luminescence analysis was carried out using TGL-68 [13]. Samples were exposed to radiation using γ-ray source ^{60}Co up to 10 kGy at temperature of liquid N_2. All measurements were carried out at 80 - 375 K, heating rate was 10 K/min. Thermostimulated luminescence data are presented in Fig.1. As follows from Fig. 1, in the case of the PMMA/DES systems, DES manifests itself as a narrow peak at 135 - 140 K (curves 2 - 4). This peak is likely to be related to glass transition of this additive in polymer matrix. Note that intensity of this peak shows a linear increase with increasing the content of DES in PMMA.

For mechanical tests, the samples were cylinders with a height of 30 mm and a diameter of 15 mm. Prior to testing, all polymer samples were annealed at temperatures above T_g by 10-15° and were slowly cooled down to room temperature.

The samples of initial PMMA and PMMA, containing DES, were uniaxially compressed at 293K with a strain rate of 0.1 mm/min using an UTS-100 tensile machine (Germany). Then, the samples were cooled with liquid nitrogen in a stressed state. Finally, the free-standing samples were heated from 213K at a heating rate of 1 K/min. The relaxation (recovery) of residual plastic deformation ε_{res} was estimated by measuring the height of the test samples with an accuracy of ±0.001 mm.

Initial structure of PMMA and PMMA with DES was studied by positron annihilation lifetime spectroscopy. Lifetime measurements were carried out using a conventional ORTEC spectrometer with a time-resolution (full width on a half maximum of the prompt coincidence curve) of 230 ps. In our experiments, ^{22}Na sandwiched between two sheets of nickel foil was used as a positron source. Annihilation in the source material, background of random coincidences and resolution function were examined using a well-known POSITRONFIT program [11].

Three-component analysis of lifetime distribution of annihilation radiation involved calculations of positron annihilation lifetimes τ_i (i =1 - 3) and intensities of the corresponding components I_i (i = 1 - 3) in measured distributions. Reliability of this examination was verified by parameter χ^2/ν with a standard deviation of 0.072. Integral statistics for the curve of delayed coincidences was (0.5 - 1.0) x 10^6 counts.

RESULTS AND DISCUSSION

Figure 2 shows the concentration dependence of yield stress σ_y for uniaxial compression of PMMA samples containing DES. As follows from Fig. 2, at 0.1 wt % of DES in PMMA, an increase in σ_y is observed. With increasing concentration of DES in polymer, σ_y decreases.

This behavior (maximum in concentration dependencies of σ_y) is quite characteristic of polymers plasticized with thermodynamically compatible low-molecular-mass compounds. Introduction of minor amounts of plasticizing agents into polymer matrix is accompanied by a concomitant increases in elastic modulus and yield stress σ_y. This phenomenon is referred to as antiplasticization. With further increase in the content of plasticizing agent, mechanical properties of plasticized polymers are much lower as compared with initial polymer. According to [14, 15], antiplasticization is rationalized by a decrease in polymer free volume and increase in polymer packing density giving rise to a dramatic inhibition of mobility of polymer chains.

Mechanical behavior of polymer containing incompatible low-molecular-mass additives is treated in terms of structural phenomena [3 - 5]. According to this approach, incompatible low-molecular-mass additive occupies structural regions with lowered packing density and increased free volume. Hence, an increase in σ_y with introduction of 0.1 wt % of DES may be related to an increase in density of these structural regions and inhibition of molecular mobility in them. As concentration of DES increases to 1.3 wt %, a decrease in σ_y is associated with loosening of these microregions and enhancement of mobility of macromolecules in them.

Experimental evidence concerning temperature-induced relaxation of plastic deformation in PMMA is presented in Fig. 3. Relaxation is seen to take place in two well-defined temperature regions: at temperatures below glass transition temperature (low-temperature component ε_1) and in the temperature region of glass transition (high-temperature component ε_2). Figure 3a shows both components for PMMA samples containing 1.3 wt % of DES.

At the present time, two alternative approaches allow explanation of this two-step character of temperature-induced relaxation of plastic deformation: gradual recovery of short-scale plastic shear transformations and excited chain conformations [16, 17] (i) and structural heterogeneity of glassy polymers [1, 2, 5] (ii) .

According to the approach concerning structural heterogeneity of glassy polymers, structure of a glassy polymer is presented as a set of structural sublevels with different packing density and ordering. Each of these sublevels is associated with its own relaxation time and activation energy of segmental mobility. Hence, at each temperature below glass transition temperature T_g, within a given sublevel, a local segmental mobility is allowed. With increasing temperature, structural sublevels with higher packing density show their segmental mobility; and, finally, at glass transition point, the whole sample is involved in segmental mobility.

In the case of the plastically deformed polymer samples, with increasing temperature, a gradual appearance of segmental mobility in structural sublevels with higher packing density is accompanied by a gradual recovery of plastic deformation via transition of excited chain conformations to initial equilibrium state within above structural sublevels. A complete recovery of plastic deformation is accomplished at T_g, when all sublevels are involved in segmental mobility.

Contribution of low-temperature component ε_1 to total recovery of plastic deformation was estimated from Fig. 3. The dependencies of contribution of ε_1 to total recovery $\varepsilon_1/(\varepsilon_1+\varepsilon_2)$ on the content of DES in PMMA are presented in Fig. 4. Comparison between Figs. 2 and 4 suggests similar changes in σ_y and $\varepsilon_1/(\varepsilon_1+\varepsilon_2)$ with increasing DES content in PMMA.

Note that examination of the effect of deformation temperature on E_o, σ_y, and $\varepsilon_1/(\varepsilon_1+\varepsilon_2)$ [1] suggests that these parameters are closely related to local segmental mobility within structural sublevels with lower packing density. Hence, the effect of concentration of DES in PMMA on σ_y and $\varepsilon_1/(\varepsilon_1+\varepsilon_2)$ is likely to be related to changes in segmental mobility in the above structural sublevels because of changes in packing density and fractional content of free volume.

To verify this assumption, the positron annihilation lifetime technique was used to follow changes on elementary free volume in polymer induced by introduction of DES into PMMA samples. Annihilation characteristics for pure PMMA and PMMA, containing DES, are summarized in Table 1. The component with lifetime τ_2 and intensity I_2 stands for preferential localization of positrons in free volume microregions within structural sublevels with an increased packing density, whereas the component with lifetime τ_3 and intensity I_3 is associated with a preferential localization of positronium in microregions involved in structural sublevels with a reduced packing density. Parameters I_2 and τ_2 are associated with concentration N_+ and effective radius R_+ of free volume microregions for the regions with increased packing density, and I_3 and τ_3 - with corresponding values for the structural sublevels with reduced packing density [6, 8].

As follows from Table I, introduction of 0.1 wt % of DES into PMMA has no effect on annihilation characteristics. As concentration of DES in PMMA increases, a slight increase in I_2 and a sharp increase in I_3 are observed. The increased values of intensities I_2 and I_3 suggest that free volume increases, i.e., introduction of low-molecular-mass additive is accompanied by structural loosening. As evidenced by more pronounced changes in I_3 as compared with I_2, this loosening is primarily associated with structural sublevels with lower packing density.

Correlation between annihilation characteristics and dimensions and concentration of free volume microregions was studied by many authors and, for example, was described in [6]. In this case, the rate of positron trapping in defects ν_+ is related to intensity I_2 of intermediate component $\tau_2 = 1/\lambda_2$ of lifetime distribution of annihilation radiation by the following relationship:

$$\nu_+ = (\lambda_f^+ - \lambda_2)I_2/[(1 - Q) - I_2] \quad (1)$$

$$\nu_+ = 4\pi D_+ R_+ N_+ \quad (2),$$

where R_+ and N_+ stand for dimensions and concentration of free volume regions occupied by positrons, and λ_f^+ is the rate of free positron annihilation.

For positronium,

$$\nu_{Ps} = (\lambda_f^+ - \lambda_3)(4I_3/3)/(Q-4I_3/3) \quad (3)$$

$$\nu_{Ps} = 4\pi D_{Ps} R_{Ps} N_{Ps} \quad (4)$$

$$\lambda_f^+ = (I'/\tau' + I_2/\tau_2 + I_3/\tau_3)/(I' + I_2 + I_3), \quad (5)$$

$$I' = 1 - I_3/3, \ \tau' = \tau_1 + [\tau_1 + \tau_3/(\lambda_s^0 \tau_3 + 1)I_3/3I',$$

where λ_f^+ is annihilation rate of free (nonlocalized) positrons, and λ_s^0 is annihilation rate of singlet positronium. Parameter Q defines the fractional content of positrons involved in positronium formation; D_+ and D_{Ps} are diffusion coefficients for positron and positronium, respectively.

Correct estimation of parameter Q is not available, and this does not allow a nonambiguous characterization. Actually, positronium atoms, which are not localized in microregions, do not manifest themselves in long-lived component with intensity I_3, and their direct identification is not possible. In connection with this, parameter Q is obtained from additional experiments or is estimated from two limiting cases. In the case of fast entrapment of all positrons by free volume microregions and existence of a certain fraction of nonlocalized positronium,

$$Q - 4I_3/3 = I_1 - I_3/3 = I'$$

and equation (3) takes the following form:

$$\nu_{Ps} = (\lambda_f^+ - \lambda_3)(4I_3/3)/I' \qquad (7),$$

where I' is associated with the fraction of nonlocalized positronium. In contrary, in the case of complete entrapment of positronium and existence of certain number of nonlocalized positrons,

$$I_1 - I_3/3 = (1 - Q) - I_2 = I'$$

and equation (1) takes the following form:

$$\nu_+ = (\lambda_f^+ - \lambda_2)I_2/I' \qquad (8).$$

Here, I' stands for the fraction of nonlocalized positrons. According to [18, 19], in the case of polymer systems, $D_+ = 0.1\ cm^2/s$ and $D_{Ps} = 10^{-4}\ cm^2/s$.

Application of these approaches allows one to estimate the effective radius R_+ and concentration of free volume regions N_+ in structural sublevels with increased packing density as well as the corresponding values of R_{Ps} and N_{Ps} for structural sublevels with decreased packing density. For all samples studied, $R_+ \simeq 0.34$ nm and $R_{Ps} \simeq 0.42$ nm. For initial PMMA and PMMA containing 0.1 wt % of DES, $N_+ \simeq 0.9 \times 10^{16}\ cm^{-3}$ and $N_{Ps} \simeq 1.0 \times 10^{19}\ cm^{-3}$. As content of DES in PMMA increases, N_+ and N_{Ps} increase. For PMMA samples containing 0.5 and 1.3 wt % of DES, $N_+ \simeq 1.15 \times 10^{16}\ cm^{-3}$ and $N_{Ps} \simeq 2.2 \times 10^{19}\ cm^{-3}$.

Hence, introduction of DES into PMMA is not accompanied with the changes in the dimensions of free volume microregions both in structural regions with increased and decreased packing density in polymer. The content of free volume microregions in more densely packed structural regions N_+ changes with increasing DES content in PMMA but still lies within expected error (20%). Concentration of free volume microregions in loosely packed structural regions N_{Ps} increases by two times, and this evidence suggests that introduction of DES into polymer is associated with loosening of these structural regions. Constancy of annihilation characteristics for PMMA samples containing 0.1 wt % of DES as compared with virgin PMMA may be rationalized by the fact that low-molecular-mass additive occupies rather large free volume microregions. These regions are located far away from each other, and distances between them are much higher than diffusion paths of positron and positronium. As a result, they appear to be beyond the detection limit of this method.

This structural interpretation of experimental evidence obtained by the method of positron annihilation correlates well with earlier structural studies in polymer glasses (for example, [20 - 22]). Structure of glassy polymer was shown to be characterized by a set of densely packed regions (domains) separated by structural regions with lower packing density. (Let us note that the term "domains" is used only to specify structural elements with increased packing density and carries no information concerning the structure of these regions.) Domains are able to aggregate with each other and produce more complicated supradomain structures. The boundary regions between supradomain structures are more loosely packed as compared with interdomain regions. Hence, experimental evidence obtained allowed identification of three structural sublevels in glassy polymers listed in order of increase in free volume: domains, interdomain regions, and boundary regions between supradomain structures. Three-component treatment of lifetime distributions of annihilation radiation allows one to obtain information concerning free volume fraction for domains and interdomain regions.

According to this approach, the effect of introduction of DES on mechanical properties of PMMA containing DES may be presented as follows. At low content of DES (0.1 wt %), DES molecules occupy rather large free volume microregions located at boundaries between supradomain structures. As a result, segmental mobility in the vicinity of these microregions is hindered, and mechanical properties of polymer material increase (Figs. 2 and 4). With increasing the content of DES in PMMA to 0.5 and 1.3 wt %, low-molecular-mass additive is able to enter interdomain structural regions. This penetration is accompanied by loosening of these regions and additional free volume microregions appear. As a result, within these structural regions, segmental mobility of macromolecules is enhanced, and σ_y and $\varepsilon_1/(\varepsilon_1+\varepsilon_2)$ decrease (Figs. 2 and 4). During introduction of DES into PMMA, densely packed domains remain intact and do not contribute to changes in mechanical properties of the system.

The effect of DES on segmental mobility of PMMA is supported by DSC data (Fig. 5). At 0.1 of DES in PMMA, endothermic jump in glass transition region increases as compared with initial PMMA (transition from curve 1 to curve 2). This evidence suggests a suppressed segmental mobility of polymer chains for polymer samples containing low-molecular-mass additive. As concentration of DES in PMMA increases (curve 3), endothermic jump decreases, and this behavior is related to an enhanced segmental mobility of polymer chains. Let us note that as evidenced by thermostimulated luminescence (Fig. 1) characterizing mobility of polymer chains and chain fragments [23], introduction of DES has a certain effect not only on large-scale (segmental) mobility of polymer chains but also on small-scale mobility of chain fragments. Introduction of 0.1 wt % of DES into PMMA is accompanied by a marked decrease in intensity of a broad peak of PMMA at temperatures from 100 to 120 K (curve 2). This behavior is related to inhibition of mobility of small kinetic elements responsible for this relaxation transitions. As concentration of DES in polymer increases (transition to curve 4), mobility of these kinetic elements is enhanced, and intensity of the peak associated with PMMA increases.

CONCLUSION

Mechanical and structural studies of PMMA containing thermodynamically incompatible low-molecular-mass additive DES allow one to explain the effect of DES on polymer mechanical behavior. At low concentration of DES in PMMA, rather large free volume microregions (presumably, at boundaries between supradomain structures) are occupied by DES molecules. As a result, segmental mobility in these regions is inhibited. This process manifests itself in increase in σ_y and contribution of low-temperature component $\varepsilon_1/(\varepsilon_1+\varepsilon_2)$ to total recovery of plastic deformation. With increasing the content of DES in PMMA to 0.5 and 1.3 wt %, low-molecular-mass agent is able to penetrate interdomain structural regions, and their structure becomes more loosened. As a result, within these structural levels, segmental mobility of polymer chains is enhanced, and σ_y and $\varepsilon_1/(\varepsilon_1+\varepsilon_2)$ decrease. During introduction of DES into PMMA, densely packed domains remain intact.

According to this approach, deformation at low strains as well as such mechanical parameters as E_o, σ_y, and $\varepsilon_1/(\varepsilon_1+\varepsilon_2)$ are controlled by segmental mobility of polymer chains in structural regions with lower packing density.

Table 1. Annihilation characteristics of PMMA containing DES for three-component analysis of positron lifetime spectra

DES Content in PMMA, wt %	Annihilation characteristics					
	τ_1	I_1	τ_2	I_2	τ_3	I_3
0	0.262 ±0.008	58.11 ±3.20	0.610 ±0.036	30.31 ±2.67	2.025 ±0.058	11.58 ±0.69
0.1	0.266 ±0.009	58.97 ±3.27	0.651 ±0.042	29.67 ±2.67	2.213 ±0.075	11.36 ±0.79
0.5	0.246 ±0.013	47.06 ±5.25	0.527 ±0.037	35.60 ±4.70	1.915 ±0.040	17.36 ±0.72
1.3	0.243 ±0.012	44.21 ±4.59	0.513 ±0.029	37.91 ±4.14	1.905 ±0.032	17.88 ±0.59

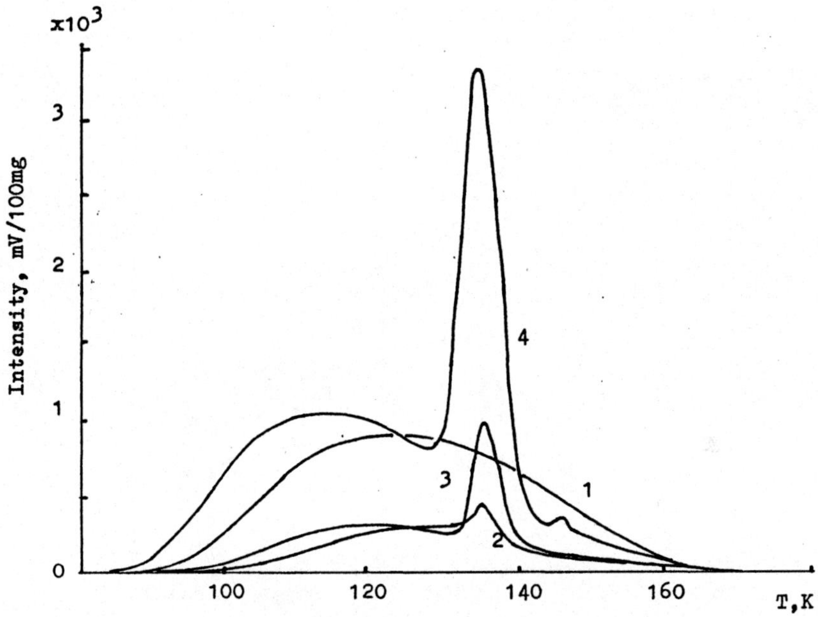

Fig.1. Thermostimulated luminiscence curves of PMMA (1) and PMMA with 0.1 (2), 0.5 (3), and 1.3 wt.% of DES (4).

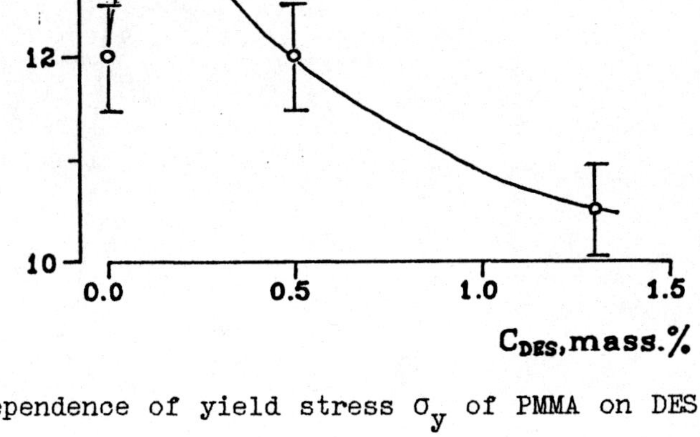

Fig.2. Dependence of yield stress σ_y of PMMA on DES content.

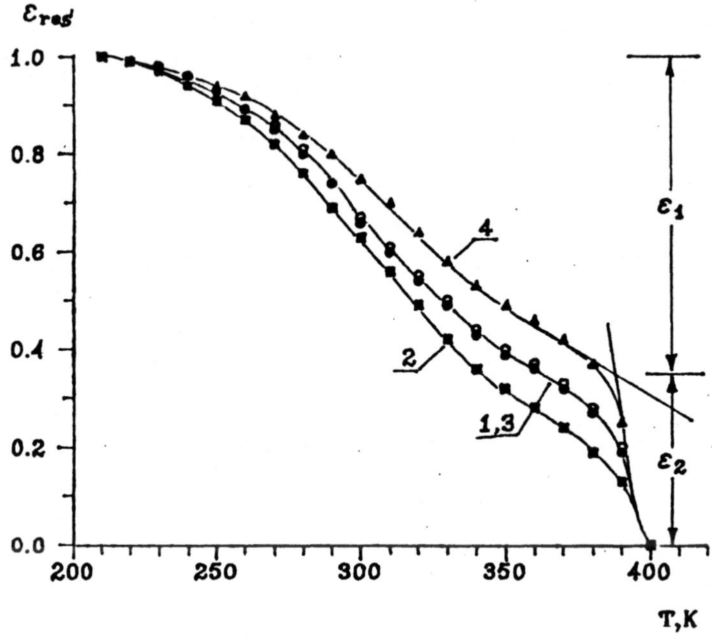
εres
1.0
0.8
0.6
0.4
0.2
0.0
200
250
300
350
400
T,K
4
2
1,3
ε1
ε2

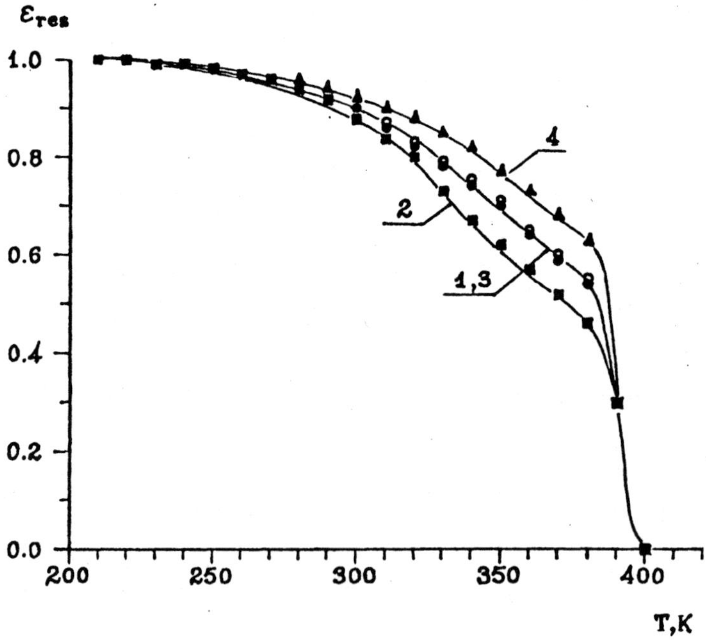
εres
1.0
0.8
0.6
0.4
0.2
0.0
200
250
300
350
400
T,K
4
2
1,3

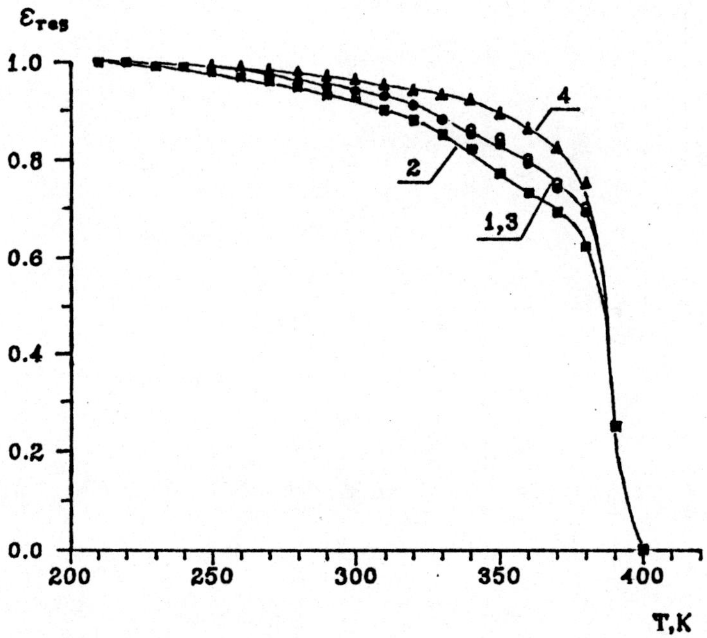

Fig.3. Recovery of residual deformation ε_{res} of PMMA (1) and PMMA with 0.1 (2), 0.5 (3), and 1.3 wt.% of DES (4). Strain: 20 (A), 30 (B), and 50 % (C).

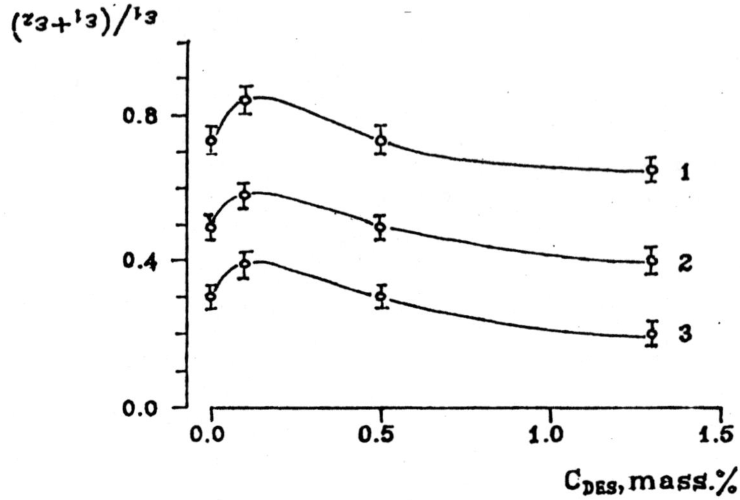

Fig.4. Contribution of low-temperature component ε_1 to total recovery of residual deformation $\varepsilon_1/(\varepsilon_1+\varepsilon_2)$ on the content of DES in PMMA. Strain: 20 (1), 30 (2), and 50 % (3).

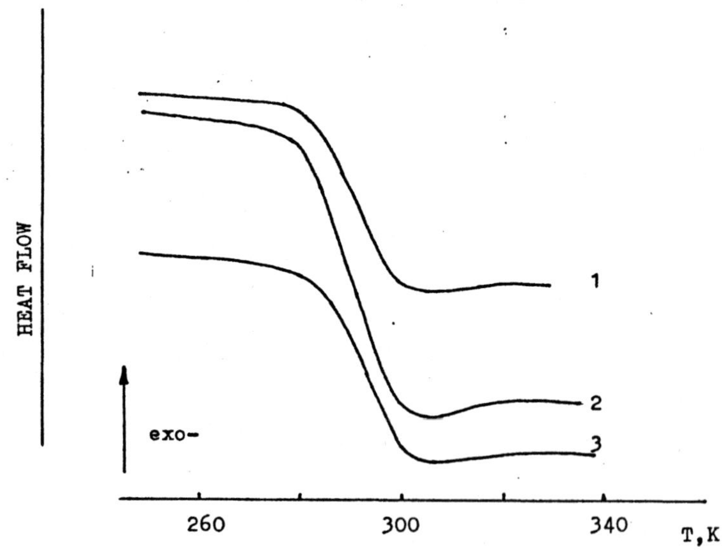

Fig.5. DSC curves of PMMA (1) and PMMA with 0.1 (2) and 1.3 wt.% of DES (3).

REFERENCES

1. M.S. Arzhakov and S.A. Arzhakov, Intern. J. Polymeric Mater., 28 (1995) (in press).

2. M.S. Arzhakov, S.A. Arzhakov, and V.A. Chernavin, New Polymeric Mater., (in press).

3. Entsiklopediya Polimerov (Encyclopedia of Polymers), Moscow: Sovetskaya Entsiklopediya, 1974 (in Russian).

4. P.V. Kozlov and S.P. Papkov, Fizikokhimicheskie Osnovy Plastifikatsii Polimerov (Physical and Chemical Foundations of Plasticization of Polymers), Moscow: Khimiya, 1982 (in Russian).

5. S.A. Arzhakov, Doctorate (Chem.) Dissertation, Moscow: 1975 (in Russian).

6. V.P. Shantarovich, Yu.P. Yampol'skii, and I.B. Kevdina, Khim. Vys. Energ., 28, 53 (1994) (in Russian).

7. I.B. Kevdina, M.S. Arzhakov, and V.P. Shantarovich, Polymer Science B, 37, 171 (1995).

8. V.P. Shantarovich, Yu.M. Sivergin, and I.B. Kevdina, Khim. Vys. Energ., (1995) (in Russian) (in press).

9. V.P. Lebedev, Usp. Khim., 47, 127 (1978) (in Russian).

10. P.H. Geil, Proc. 17th Int. Symp., 83 (1987).

11. V. Sanchez, R. Lopez, L.A. Fucugauchi, and Y. Ito, J. Appl. Polym. Sci., 56, 779 (1995).

12. R.C. Macquenn and R.D. Granata, J. Polym. Sci. B, 31, 971 (1993).

13. V.G. Nikol'skii and N.A. Mironov, Zavodskaya Laboratoriya, 1272 (1973) (in Russian).

14. J.S. Vrentas, J.L. Duda, and H.-C. Ling, Macromolecules, 21, 1470 (1988).

15. Y. Liu, A.K. Roy, A.A. Jones, P.T. Inglefield, and P. Ogden, Macromolecules, 23, 968 (1990).

16. E.F. Oleinik, O.B. Salamatina, S.N. Rudnev, and S.V. Shenogin, Polymer Science, 35, 1532 (1993).

17. E.F. Oleinik, O.B. Salamatina, S.N. Rudnev, and S.V. Shenogin, Polymers for Advanced Technologies, 6, 1 (1995).

18. M. Maurino and W. Brandt, Bull. Am. Phys. Soc., 24, 72 (1979).

19. M. Eldrup, A. Vehanen, P.J. Schulz, and K.G. Lynn, Phys. Rev. Lett., 51, 2007 (1983).

20. K. Neki and P.H. Geil, J. Macromol. Sci. B, 8, 295 (1973).

21. H.E. Bair, M. Matsuo, W.A. Salmon, and T.K. Kwei, Macromolecules, 5, 114 (1972).

22. J.L. Racich and J.A. Koutsky, J. Appl. Polym. Sci., 20, 2111 (1976).

23. V.G. Nikol'skii, Pure Appl. Chem., 54, 493 (1982).

Physical and Mechanical Behavior of Polymer Glasses I. Polymeric Powders

M.S. Arzhakov and S.A. Arzhakov

M.S. Arzhakov and S.A. Arzhakov

Polymer Department, Faculty of Chemistry, Lomonosov Moscow State University
Vorob'evy Gory, Moscow, 119899 Russia

Deformation behavior of glassy thermoplastic powders under the parallel action of pressure and temperature was studied. Three "pressure-temperature" regions of preparation of monolithic polymer samples with quite different physical and mechanical properties were found. The existence of these regions was shown to be controlled by the relaxation in powderlike polymer during compression. Speculations concerning segmental mobility within polymer glassy state were proposed to describe specific relaxation behavior in the bulk polymer samples.

Key Words: polymer glasses, powders, molding, deformation, relaxation

INTRODUCTION

At the present time, a variety of polymer thermoplastics are produced as powders and granules. They are processed into final monolithic articles by pressing, extrusion, injection molding, *etc.* [1-4]. Physical and mechanical properties of the final bulk materials are primarily controlled by deformation and relaxation in powderlike polymers during processing. Hence, studying deformation behavior of polymer powders under the parallel action of pressure and temperature is of a great importance from the standpoint of the fundamental approaches to polymer processing.

On the other hand, the problems of molding of polymeric powders are closely related to the fundamentals of mechanical behavior of polymers under high pressure. These problems are widely discussed in literature [5-9]. However, in all cases, physical and mechanical behavior of polymers under high pressure was studied for bulk samples, whose pre-history is controlled by processing conditions.

The aim of this work is a thorough description of deformation of powderlike glassy thermoplastics under the parallel action of temperature and pressure, which results in monolithization, and its governing effect on the physical and mechanical behavior of bulk polymers.

EXPERIMENTAL

In this work, we studied PMMA, crosslinked PMMA, copolymers of methyl methacrylate with methacrylic acid, PVC, PS, poly (methacrylic acid) (PMAA), polyacrylonitrile (PAN), and copolymers of acrylonitrile with vinyl acetate.

PMMA (M_w=1,500,000) as well as copolymers of methyl methacrylate with methacrylic acid were prepared by bulk polymerization. The content of methacrylic acid in copolymers was 15 and 25 wt.%. Copolymers of methyl methacrylate with bifunctional monomer, ethylene glycol dimethacrylate (EGDM), were used as cross-linked PMMA. The development of network took place immediately during bulk copolymerization. The content of EGDM in polymerization system was 1.5 and 8.0 mol.%.

PVC (M_w=79,000) was prepared by bulk polymerization.

Bulk polymers and copolymers were dispersed to powders, the dimensions of particles do not exceed 0.25 mm.

Dibutyl phthalate (DBPh) was used as a plasticizing agent for PMMA and PAN.

Molding of polymer powders was carried out under two regimes. A certain amount of powderlike polymer was placed in cylindrical mold with a diameter of 10 mm and compressed at room temperature T_r with a strain rate of 0.1 mm/min up to a certain stress. Then, the stressed sample was heated up to molding temperature T_{mold}, allowed to stay for 15 min, cooled down to T_r, and released ("P-T" regime). In "T-P" regime, a certain amount of polymer powder was heated in the mold up to T_{mold}, compressed with a strain rate of 0.1 mm/min up to a certain stress, allowed to stay for 15 min, cooled down to T_r, and released.

Monolithization of the molded samples (the absence of visual boundaries between the particles of initial powder) was visual inspected at T_r, as evidenced by their transparency. Accuracy of estimation of temperature and pressure of transition from opaque samples to transparent was 5-7%.

To study deformation behavior of powderlike polymers, a certain amount of polymeric powder was compacted in the mold at 0.03 GPa at T_r for 15 min. Then, the samples were unloaded, and initial height of the test samples h_o was estimated. After that, the samples were compressed at a strain rate of 0.1 mm/min up to a certain stress. At this stress, the height of the deformed sample was h_p. Strain was estimated as $\Delta h=h_o-h_p$. The accuracy of measurements was 3%. Deformation of powderlike polymers was carried out in the temperature range from 293 K to glass transition temperature T_g.

To study the temperature-induced relaxation of deformation, the samples molded at certain temperatures and pressures were cooled in the stressed state to the temperature of liquid nitrogen and were released at this temperature. Then, the free-standing samples were heated with a heating rate of 1 K/min starting with 213 K, and the changes in their height were measured with an accuracy of 0.001 mm. The changes in the height of the test samples during their heating was estimated as $\Delta h=h_T-h_{213}$, where h_{213} is the height of the sample at 213 K, and h_T is the height of the sample at a given temperature.

RESULTS AND DISCUSSION

Molding of PMMA, PVC, and PS powders under "P-T" regime showed that the transparent monolithic samples with no boundaries between particles of initial powder may be prepared in a certain region of temperatures and pressures. For PMMA, this region is bounded by the curve presented in Fig. 1.

Compression of powderlike polymer at T_r within the pressure interval AB (Fig. 1), which is followed by heating to T_{mold} (in the case presented in Fig. 1, T_{mold} = 400 K), leads to the formation of opaque samples. Transparent samples may be prepared within interval of pressures BC. Further increase in pressure again results in the formation of opaque samples.

The most interesting feature of this experimental evidence is related to the existence of the upper branch of the curve bounded the region of preparation of transparent samples. This fact was first reported in [10]. To study this phenomenon, let us consider the deformation of PMMA powder in the mold at T_r.

Pressure-strain curve may be divided into three regions (Fig. 2). Within region I ($P \le P_1$), the dependence of pressure on strain shows a linear character. This region is characterized by the densification of powderlike system due to the Hookean elastic deformation of particles of initial powder. As evidenced by optical microscopy, the shape of particles remains unchanged after deformation. Within region II (from P_1 to P_2), densification proceeds via yielding of particles. At $P \ge P_2$, compression of monolithic sample takes place. Let us note that the compressibility factor of powderlike polymer calculated within this region is equal to that observed for the deformation of bulk PMMA under the same conditions.

With increasing deformation temperature, a decrease in the pressures associated with the transition from Hookean elastic deformation to yielding (P_1) and from yielding to the deformation of monolithic body (P_2) is observed. Figure 3 shows the temperature dependencies of P_1 and P_2 (curves 1 and 2, respectively) as well as the curve, which bounds the region providing preparation of transparent samples (curve 3).

At temperatures and pressures corresponding to region I (Fig.3), Hookean elastic deformation of powderlike polymer takes place. At pressures above P_1 (Fig.3, curve 1), within region II, the densification of polymer powder is provided by yielding. At pressures above P_2 (Fig.3, curve 2), mechanical behavior of polymer powder is similar to that of bulk PMMA. Hence, one may conclude that, under these conditions, the boundaries between powder particles disappear. To verify this assumption, polymer samples compressed in region IV were cooled down in mold in the stressed state with liquid nitrogen and released at this temperature. At the temperature of liquid nitrogen, the samples are transparent but they become opaque on heating to T_r. Hence, at temperatures and pressures corresponding to region IV, powderlike polymer is monolithic under compression.

To summarize, monolithization of polymer powder (disappearance of boundaries between powder particles) takes place within regions III and IV. Generally, monolithization is provided by the diffusion of macromolecules or their fragments through the boundaries between particles. Let us note that the samples prepared in region IV are monolithic at rather low temperatures below T_r. In contrast, the samples prepared in region III remain monolithic at T_r.

For the samples molded within region IV on heating to T_r the loss in transparency is explained by polymer relaxation associated with the recovery of the boundaries between powder particles. These processes are not related to the relaxation of Hookean deformation, because, at low temperatures, the samples remain monolithic for a long period of time. From this standpoint, at a given pressure, the temperature corresponding to the relaxation of the component of relaxation spectrum that is observed in molded samples below T_r is described by an upper branch of the curve 3 (Fig. 3). Increase in macromolecular mobility provided by plasticization enhances relaxation and, finally, leads to shifting region III toward lower temperatures (Fig. 4).

When molding is carried out in "T-P" regime, the upper branch of the curve, which bounds the area of formation of transparent samples, is shifted toward lower temperatures (Fig.5). This behavior may be rationalized by the fact that in this regime powderlike polymer is compressed at temperatures above T_r. It leads to the development of relaxation during compression. It is worth noting that increase in rate of compression leads to shift of upper branch toward that obtained in "P-T" regime.

Studies in molding of PAN and PMAA powders using "P-T" regime showed the absence of upper pressure limit of preparation of transparent samples (Fig.6). It evidences that these polymers are characterized by the absence of relaxation at $T < T_r$. This fact is likely to be connected to high energy of intermolecular interaction specific for PAN and PMAA.

Decrease in energy of intermolecular interaction of PAN due to plasticization or copolymerization with vinyl acetate results in appearance of upper branch of P-T curve (Fig.7). The same behavior is characteristic for copolymers of methyl methacrylate with methacrylic acid (Fig.8). On contrary, introduction in polymer intermolecular covalent bonds in the case of copolymers of methyl methacrylate with EGDM is accompanied by the restriction of relaxation and shift of the region where transparent samples can be formed toward higher temperatures (Fig. 9).

It is worth to emphasize that the preparation of the samples which are transparent and monolithic at T_r does not unambiguously evidence the temperature stability of polymer material obtained. To understand in detail deformation and relaxation behavior of powderlike polymer under molding let us consider temperature-induced relaxation of deformation of molded samples.

Temperature dependences of the height of PMMA molded samples are presented in Fig.10. A noticeable portion of deformation of PMMA samples molded in region IV (Fig.3) relaxes at temperatures below T_g (Fig.10 A, curve 1). Increase in height of molded sample during heating is controlled by the relaxation of deformation of powder particles. It leads to the recovery of the boundaries between powder particles and, as a result, loss in monolithization.

As was mentioned earlier, monolithization within region IV is controlled by yielding of powder particles, that is, the diffusion of segments of macromolecules through the boundaries between powder particles. Hence, temperature-induced relaxation of samples prepared in region IV appears to be also controlled by segmental mobility. Let us note, that, in this case, the appearance of segmental mobility takes place at temperatures below T_g.

To explain this phenomenon, let us consider the mechanism of yielding [4, 11-13]. Yielding is controlled by the mechanical activation of segmental mobility in polymer within glassy state. This process takes place when relaxation time of segments $\tau = \tau_0 e^{(U-\gamma\sigma)/kT}$, where U is the energy of activation of segmental mobility, becomes equal to the time of strain development due to a decrease in U under the applied stress σ. The mechanism of low-temperature segmental mobility in the molded samples is supposed to be similar to that. However, in this case, mechanical activation of segmental mobility is observed not under applied stress but under inner stress stored in the deformed polymer sample.

Increase in T_{mold} leads to degeneration of low-temperature relaxation and preparation of the samples, which show relaxation only in the vicinity of T_g (Fig. 10 A, curves 2,3). Let us note that degeneration of low-temperature relaxation takes place under the transition from region IV to region III (Fig. 3).

Hence, in general case, temperature-induced recovery of the dimensions of the molded polymer samples is controlled by the relaxation of deformation of particles of initial polymer powder and involves low-temperature component within polymer glassy state and high-temperature component in the vicinity of T_g. Both components are related to the appearance of segmental mobility in the deformed polymer samples. At temperatures below T_g, segmental mobility is induced under inner stresses stored in polymer during molding. With increasing T_{mold} degeneration of low-temperature component of relaxation controls the critical temperature and pressure of transition from region IV to region III (Fig. 3, upper branch of the curve 3). These critical values determine the temperature of relaxation of the low-temperature component of segmental mobility spectrum at a given pressure during compression. Increase in T_{mold} above this temperature leads to the preparation of polymer samples, which show relaxation only at T_g. This reasoning allows one to conclude that the upper branch of curve 3 (Fig. 3) is the pressure dependence of T_g.

Relaxation of deformation of the samples molded in region III (Fig.10 B, C) suggests, that increase in T_{mold} is accompanied by degeneration of high-temperature relaxation in the vicinity of T_g. When the critical T_{mold} is achieved, the changes in the height of the molded samples is controlled only by thermal expansion (Fig.10 B, C, curves 3). This evidence implies, that, under these conditions, true monolithization takes place.

The results obtained show the existence of several "pressure-temperature" regions providing the preparation of monolithic samples with quite different properties. The general "pressure-temperature" diagram for PMMA is presented in Fig. 11.

Preparation of the samples, which remain monolithic at temperatures above T_g, is allowed in region I. In this case, the monolithization of powderlike polymer is controlled by the diffusion of centers of masses of macromolecules through the boundaries between the powder particles. The lower branch of the curve bounded region I is associated with both temperatures and pressures, at which diffusion of centers of masses of macromolecules through the boundaries between the powder particles is allowed. The existence of the upper branch of the above curve is related to the fact that diffusion of the macromolecules requires the existence of the sufficiently large free volume microregions in polymer. With increasing pressure, these free volume microregions are likely to be dispersed, and their sizes decrease. As a result, diffusion of macromolecules is prohibited. From this standpoint, the upper branch of the curve bounding region I represents the pressure dependence of flow temperature. Let us note that for PMMA cross-linked with EGDM, when diffusion of macromolecules is not allowed, region I is not observed.

Within regions II and III, monolithization of polymer powder during molding also takes place. In these regions, monolithization is controlled by the diffusion of segments through the boundaries between the powder particles. This process is accompanied by the development of noticeable inner stresses, which are likely to be stored in polymer sample after release. On heating the samples molded in region III, segmental mobility is observed even at temperatures below T_g. This process leads to the recovery of the boundaries between powder particles and loss of monolithization. Molding in region II is accompanied by relaxation of low-temperature component of segmental spectrum during compression. Hence, in this case, segmental mobility in the molded sample takes place only at T_g. The loss in monolithization is observed only in this temperature interval.

At temperatures and pressures corresponding to region IV, opaque samples are produced.

CONCLUSION

Examination of molding of powders of glassy thermoplastics under the parallel action of temperature and pressure showed the existence of three "pressure-temperature" regions providing the preparation of monolithic polymer samples.

Preparation of true monolithic samples is observed in the "pressure-temperature" region, where diffusion of macromolecules through the boundaries between the particles of initial powder takes place. When monolithization is provided by the diffusion of segments through the boundaries between the powder particles, the resultant physical and mechanical behavior of bulk polymer samples is controlled by the segmental relaxation both during molding and after release of molded polymer.

In general case, relaxation of yield deformation of powderlike polymer involves low-temperature component within polymer glassy state and high-temperature component in the vicinity of T_g. Both components are related to the appearance of segmental mobility in the deformed glassy polymers on heating. At temperatures below T_g, segmental mobility is provided by inner stresses stored in polymer samples during their molding.

Evidently, these features of relaxation of yield deformation in glassy polymers are beyond the scope of molding of polymer powders. Consideration of low-temperature relaxation of plastically deformed glassy polymers seems to be of primarily importance because it is not practical to prepare polymer articles without inner stresses. Relaxation behavior of the plastically deformed bulk polymers will be discussed in the forthcoming papers.

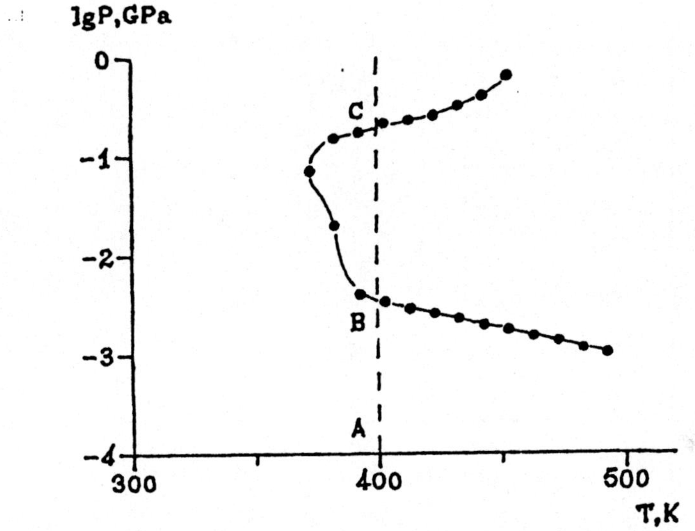

Fig. 1. "Pressure-temperature" diagram for the preparation of transparent PMMA samples.

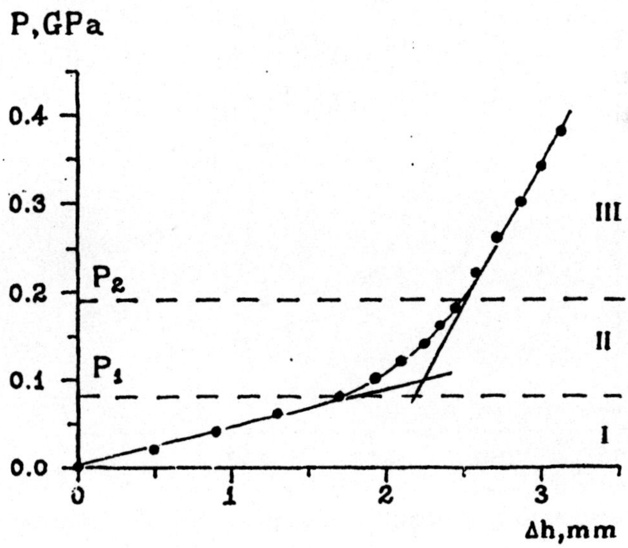

Fig. 2. Pressure-strain curve of powderlike PMMA.

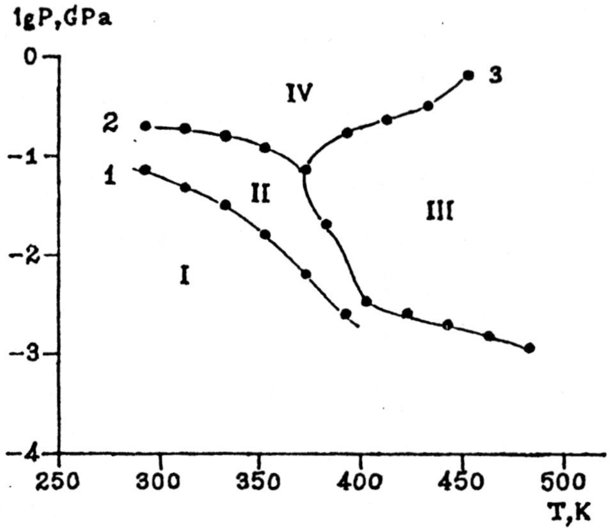

Fig. 3. Temperature dependences of pressure P_1 (1) and P_2 (2). 3 - curve bounding the region of preparation of transparent PMMA samples.

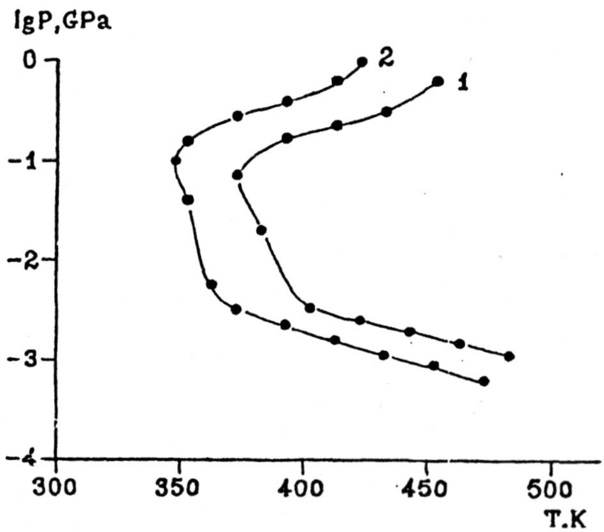

Fig. 4. "Pressure-temperature" diagram for the preparation of transparent samples from PMMA (1) and PMMA plasticized with DBPh (2). (Content of DBPh: 5 wt.%).

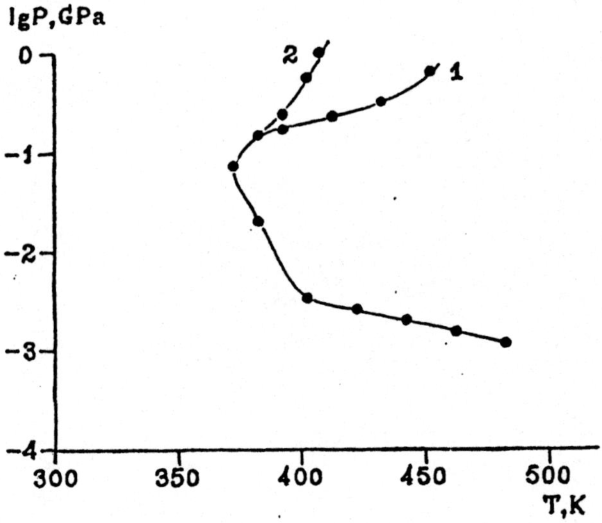

Fig. 5. "Pressure-temperature" diagram for the preparation of transparent PMMA samples under "P-T" (1) and "T-P" regimes (2).

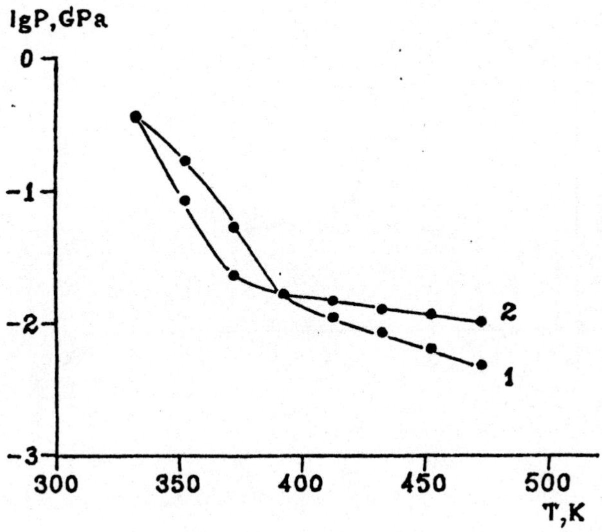

Fig. 6. "Pressure-temperature" diagram for the preparation of transparent samples of PAN (1) and PMAA (2).

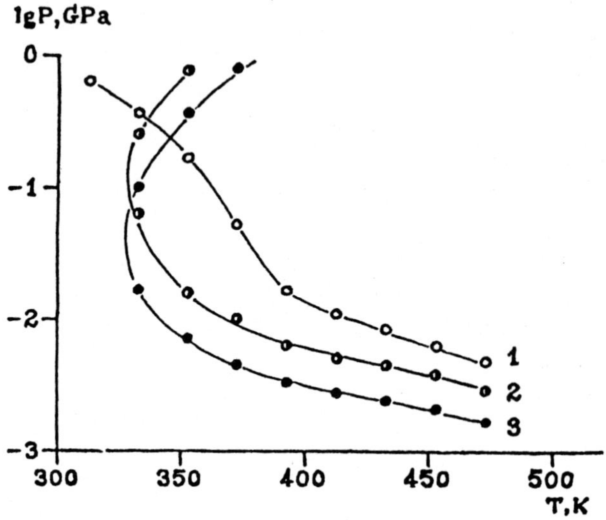

FIG. 7 A

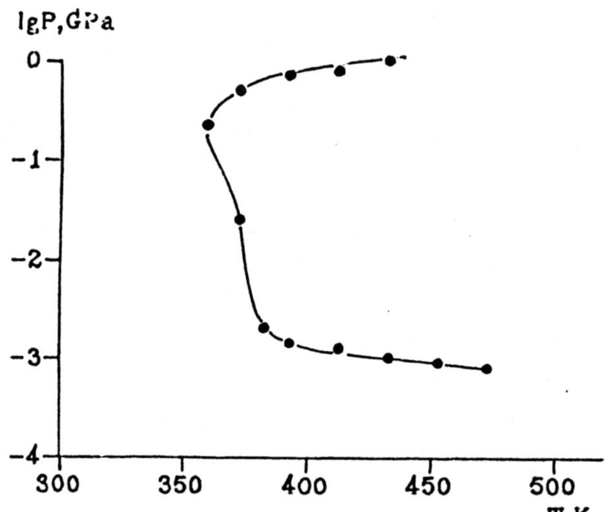

FIG. 7 B

Fig. 7. "Pressure-temperature" diagram for the preparation of transparent samples from PAN (1) and PAN plasticized with 9 (2) and 14 wt.% DBPh (3) (A), and from copolymer of acrylonitrile with 15% of vinyl acetate (B).

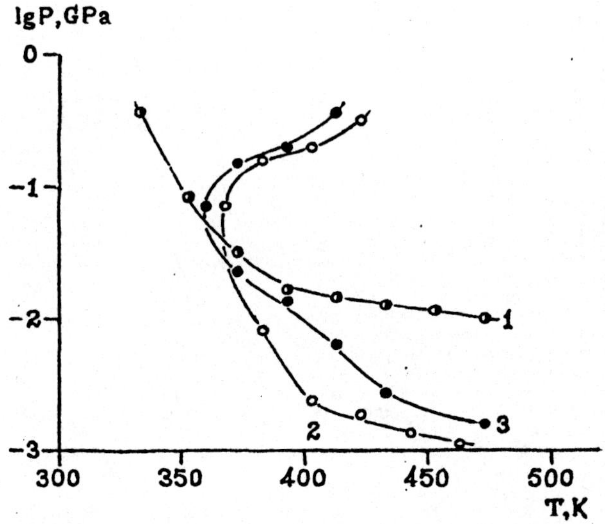

Fig. 8. "Pressure-temperature" diagram for the preparation of transparent samples from PMAA (1) and copolymers of methacrylic acid with 15 (2) and 25 wt.% of methyl methacrylate (3).

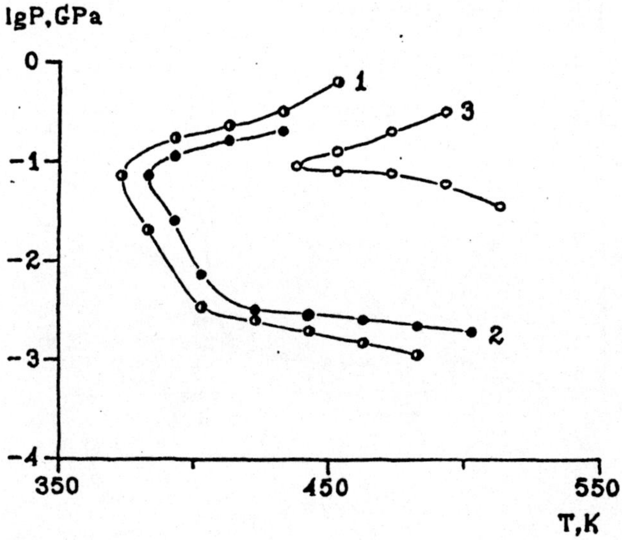

Fig. 9. "Pressure-temperature" diagram for the preparation of transparent samples from PMMA (1) and copolymers of methyl methacrylate with 1.5 (2) and 8 mol.% of EGDM (3).

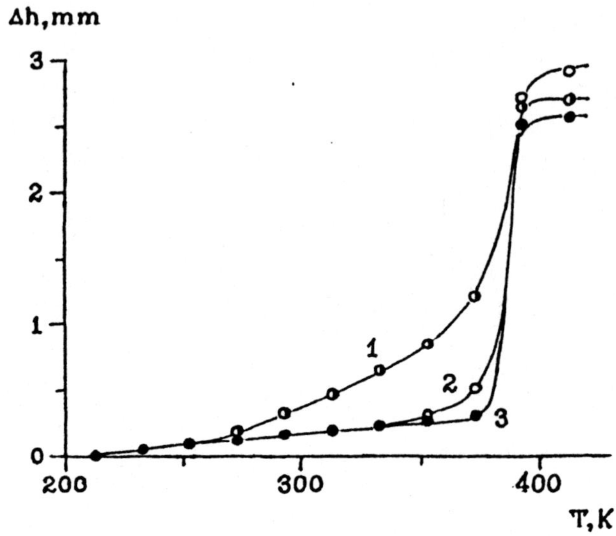

FIG. 10 A

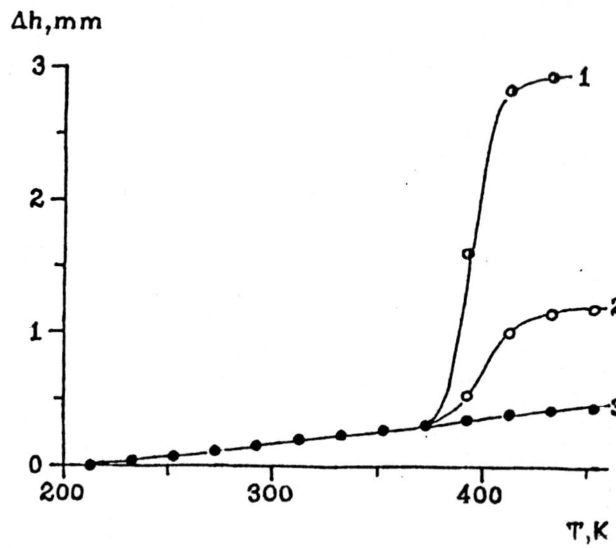

FIG. 10 B

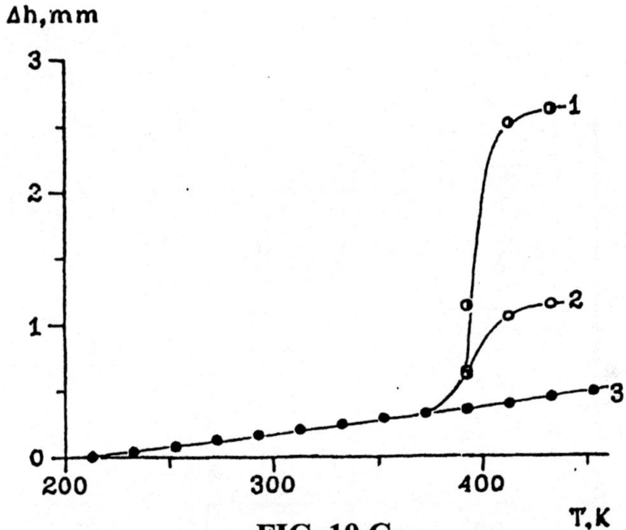

FIG. 10 C

Fig. 10. Temperature dependences of the height of PMMA samples molded at 0.15 GPa and temperatures 353 (1), 393 (2), and 433K (3) (A); at pressure 0.04 GPa and temperatures 393 (1), 433 (2), and 473K (3) (B); at pressure 0.02 GPa and temperatures 393 (1), 433 (2), and 453K (3) (C).

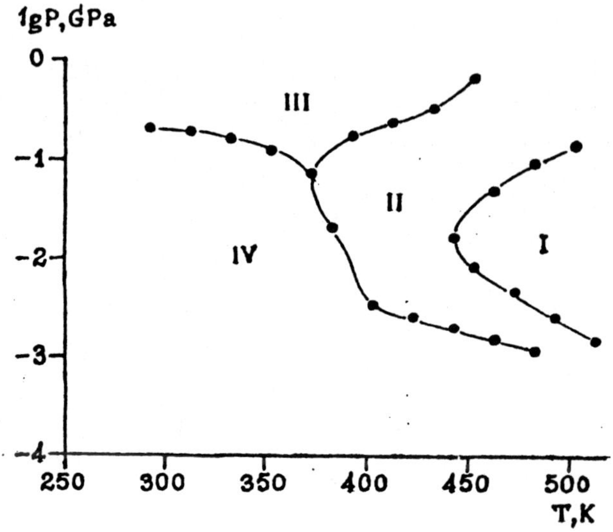

Fig. 11. General "pressure-temperature" diagram for molding of powderlike PMMA. Comments are given in text.

REFERENCES

1. Z. Tadmor and K. Gogos, Teoreticheskie Osnovy Pererabotki Polimerov (Theoretical Foundations of Polymer Processing), Moscow: Khimiya, 1984 (in Russian).

2. V.E. Gul' and M.S. Akutin, Osnovy Tekhnologii Pererabotki Plastmass (Foundations of Technology of Plastics Processing), Moscow: Khimiya, 1985 (in Russian).

3. V.R. Gowariker, N.V. Viswanathan, and J. Sreedhar, Polymer Science, New Delhi: Wiley, 1986.

4. Entsiklopediya Polimerov (Encyclopedia of Polymers), Moscow: Sovetskaya Entsiklopedia, 1977 (in Russian).

5. V.F. Skorodumov and Yu.K. Godovskii, Vysokomol. Soedin., 35, 214 (1993) (in Russian).

6. J. Havlicek, M. Ylalsty, and J. Hrouz, J. Macromol. Sci., Phys., 21, 425 (1982).

7. J.A. Sauer, Polym. Eng. Sci., 17, 150 (1977).

8. D.S. Sanditov and G.M. Bartenev, Vysokomol. Soedin., Ser. A, 32, 848 (1990) (in Russian).

9. S.B. Ainbinder, K.I. Alkne, E.P. Tyunina, and M.G. Laka, Svoistva Polimerov pri Vysokikh Davleniyakh (Properties of Polymers at High Pressures), Moscow: Khimiya, 1973 (in Russian).

10. S.A. Arzhakov, Doctorate (Chem.) Dissertation, Moscow: Karpov Institute of Physical Chemistry, 1975 (in Russian).

11. The Physics of Glassy Polymers, R.N. Haward, Ed., New York, Toronto: Wiley, 1973.

12. L.E. Nielsen, Mechanical Properties of Polymers and Composites, New York: Dekker, 1974.

13. V.A. Kargin, Struktura i Mekhanicheskie Svoistva Polimerov (Structure and Mechanical Properties of Polymers), Moscow: Nauka, 1979.

Physical and Mechanical Behavior of Polymer Glasses II. Poly(Methyl Methacrylate)

M.S. Arzhakov and S.A. Arzhakov

Polymer Department, Faculty of Chemistry, Lomonosov Moscow State University, Vorob'evy Gory, Moscow, 119899 Russia

For PMMA samples under compression in the press mold, physical and mechanical behavior was characterized. The development of plastic deformation and accumulation of residual deformation were shown to take place even at temperatures far below glass transition temperature. The pressure dependence of glass transition temperature as well as the compression stress-strain diagram of PMMA samples showed well-pronounced inflection points at pressures of 0.09 - 0.1 and 0.38 - 0.4 GPa. Physical and mechanical behavior of PMMA was discussed in terms of structural pattern of plastic deformation concerning the gradual involvement of structural sublevels with different packing densities in deformation and relaxation.

Key words: PMMA, plastic deformation, relaxation, structure

INTRODUCTION

Temperature-induced relaxation of plastic deformation in polymer glasses is known [1 - 4] to involve low-temperature component within glassy state and high-temperature component at temperatures close to glass transition temperature T_g. The high-temperature component is associated with the transition of excited chain conformations to their initial equilibrium state via segmental mobility of polymer chains. The origin of the low-temperature component of relaxation of residual plastic deformation ε_{res} is not so clear. At the present time, the appearance of this mode of relaxation is treated in terms of two approaches: recovery of small-scale plastic shear transformations [3, 4] (i) and structural inhomogeneity of glassy polymers [2, 5, 6] (ii). Within the framework of the latter approach, structure of glassy polymer involves a set of structural sublevels with different packing densities and degrees of ordering.

Each of these structural sublevels is characterized by its own relaxation time and activation energy of segmental mobility. Hence, at each temperature below T_g, local segmental mobility in certain structural sublevels is allowed. With increasing temperature, structural sublevels with higher packing densities acquire segmental mobility, and finally at T_g, all polymer sample is involved in segmental mobility. Hence, a wide spectrum of segmental mobility in glassy polymer is provided by the existence of a set of structural sublevels with different packing densities.

As shown earlier [5, 6], there is a good correlation between the contribution of low-temperature component to the temperature-induced relaxation of ε_{res} and mechanical parameters of uniaxial compression of glassy polymer. This experimental evidence allowed one to suggest that the deformation of polymer at strains below yield strain ε_y is developed via deformation of structural sublevels with lower packing densities. In the case of uniaxial compression, these speculations were supported by structural studies of deformation of PMMA samples using the method of positron annihilation lifetime spectroscopy [7 - 9].

The results obtained allowed us to advance the following structural pattern of the development of plastic deformation in glassy polymers. At strains below ε_y, deformation is controlled by the stress-induced segmental mobility within the structural sublevels with lower packing densities. At temperatures well below T_g, temperature-induced relaxation of this portion of plastic deformation is controlled by low-temperature portion of the spectrum of segmental mobility, which is provided by gradual appearance of segmental mobility in the above structural sublevels. At strains above ε_y, plastic deformation is controlled by the stress-induced segmental mobility within the densely packed structural sublevels. This portion of plastic deformation is able to relax at temperatures close to T_g when all polymer sample is involved in segmental mobility.

Obviously, structural inhomogeneity of polymer glasses, that is, inhomogeneity of packing densities within different structural sublevels, seems to be even more pronounced under the triaxial compression of polymers. Uniaxial compression is accompanied by increase in lateral sizes of the polymer sample, and plastic deformation shows rather complicated character. To gain a detailed information concerning the effect of structural inhomogeneity on the development of plastic deformation, one should invoke such mode of deformation when compression-induced decrease in the height of the sample is accompanied by no changes in lateral direction. This particular mode of loading may be realized by compression of polymer samples in cylindrical press mold. Earlier [10], this mode of compression was applied to investigate the monolithization of powderlike polymer thermoplastics.

Note that, the studies concerning the deformation of powderlike glassy thermoplastics in the press mold under the parallel action of pressure and temperature evidenced that temperature-induced relaxation of plastic deformation of powderlike polymer also involves low-temperature component and high-temperature component. Both components are associated with the appearance of segmental mobility in deformed polymer sample on heating. The mechanism of the low-temperature component of relaxation is supposed to be controlled by the activation of low-temperature segmental mobility induced by the stored inner stress in the deformed polymer sample. However, the complicated character of deformation of powderlike polymers under their compression in the press mold does not allow one to obtain a detailed information concerning the deformation behavior of glassy polymers at this particular mode of loading.

The aim of this work was to study the deformation of bulk

PMMA in the press mold to reveal the effect of structural inhomogeneity of glassy polymers on their physical and mechanical behavior.

EXPERIMENTAL

In this work, PMMA prepared by bulk polymerization (M_η = $3x10^6$) was used. The test samples were cylinders with a height of 20 mm. Prior to testing, polymer samples were annealed at temperatures above T_g by 10 - 15 K, and then they were slowly cooled down to room temperature.

To study the temperature-induced relaxation of plastic deformation, polymer samples were compressed in the cylindrical press mold with a diameter of 10 mm up to pressures 0.3 and 0.5 GPa with a strain rate of 0.1 mm/min. Deformation temperatures T_{def} were varied from 173 to 463 K. The tests were performed using an UTS-100 tensile machine (Germany). To provide the close contact between sample and press mold, the following procedure was used. The difference between thermal expansion of the sample and the press mold was accounted for. When tests were performed at low temperatures, the sample with a certain diameter was placed in the press mold at temperatures by 10 - 20 K below T_{def}. As temperature increases to T_{def}, a close contact between the sample and the press mold was provided as a result of thermal expansion of polymer. When tests were performed at elevated temperatures, the sample with the certain diameter was placed in the press mold at room temperature. As temperature increases to T_{def} the close contact between sample and press mold was provided as a result of thermal expansion of polymer.

Note that, reference hydrostatic compression of bulk PMMA was carried out. These tests demonstrated a fair agreement between the changes in the volume of the samples compressed in the press mold and under hydrostatic loading in the studied range of pressures and temperatures. This experimental evidence allowed one to conclude that compression in the mold does not involve uniaxial deformation because of the clearance between mold and sample.

After loading of the samples up to a certain pressure they were kept at this pressure until their equilibrium height was attained. Then, the stressed samples were cooled down with liquid nitrogen and unloaded at this temperature. The free-standing samples were heated from 213 K at a heating rate of 1 K/min. The relaxation (recovery) of residual deformation was estimated by measuring the height of the samples with an accuracy of ± 0.001 mm.

To obtain the pressure dependence of T_g, the isothermal compressibility of PMMA was studied at pressures varying from 0.03 to 0.66 GPA at temperatures from 370 to 490 K. To provide a close contact between the sample and the press mold, the samples in the press mold were heated up to 500 K. Then, the samples were loaded to a pressure of 0.015 GPa and kept at this pressure for 15 min. Then, the stressed samples were cooled down in the press mold to the test temperature with a rate of 0.2 K/min. At test temperature, the samples were kept at this pressure until the equilibrium height was attained. After this procedure the samples were unloaded and kept at test temperature until the equilibrium height h_o was attained. No changes in the weight of the samples, molecular mass of polymer, and concentration of residual monomer in PMMA evidenced that during this training physical and chemical changes in polymer did not take place.

At the test temperature, the samples were compressed up to a certain pressure with a loading rate of 0.1 mm/min. The stressed samples were kept at this pressure until the equilibrium height h_P was attained. Then the pressure was increased. This experimental evidence allowed one to estimate the specific equilibrium volume of polymer at a given pressure and temperature. Thermal expansion and compressibility of working parts of tensile machine were taken into account. Specific volume was estimated with an accuracy of ± 0.25%.

The pressure dependence of specific volume of polymer at a given temperature allowed one to estimate the isothermal compressibility factor β, that is, the relative change in volume with increasing pressure by ΔP:

$\beta = \dfrac{\Delta V}{V_o \Delta P}$, where $\Delta V = V_o - V_P$, V_o is the specific volume of the polymer before loading, V_P is the specific volume of the polymer at pressure P. Temperature dependences of β at different pressures are shown in Fig. 1.

The deformation behavior of PMMA under the compression in the press mold was studied at 293 K. Before tests, the samples were treated according to the procedure similar to that used for the studies of isothermal compressibility of PMMA samples. The equilibrium height of the as-prepared sample was h_o. The samples were loaded in the press mold up to a certain pressure with a loading rate of 0.1 mm/min. At this pressure, the samples were kept until the equilibrium height h_p was attained. Then, the pressure was again increased up to a certain value. Deformation of polymer was estimated as $\varepsilon = (h_o - h_p)/h_o$.

RESULTS AND DISCUSSION

Figure 2 shows the temperature dependences of residual deformation ε_{res} for PMMA samples compressed in the press mold in the wide range of temperatures. For the samples compressed at temperatures below glass transition temperature T_g, the recovery of ε_{res} takes place within the glassy state via low-temperature component of relaxation (Fig. 2 A, curves 1 - 5; Fig. 2 B, curve 1). A complete relaxation of ε_{res} is observed at temperatures close to deformation temperature T_{def}. Hence, at temperatures far below T_g (the lowest T_{def} was 173 K), compression of PMMA samples in the press mold is accompanied by development of plastic deformation and accumulation of ε_{res}.

Note that, for the samples compressed at the temperature of liquid nitrogen (78 K), the complete relaxation takes place during unloading at this temperature. This experimental evidence allows one to conclude that, in this case, only elastic Hookean deformation occurs. Critical temperature corresponding to the transition from elastic Hookean deformation to plastic deformation lies in the temperature range from 78 to 173 K.

According to the structural pattern of plastic deformation

of polymer glasses [5, 6], the development of plastic deformation of polymer under compression in the press mold is likely to be associated with the stress-induced segmental mobility within the structural sublevels with lower packing densities. On heating unloaded samples, temperature-induced segmental mobility within above structural sublevels results in relaxation of ε_{res} within glassy state.

For PMMA samples compressed at temperature much above T_g, relaxation of ε_{res} is observed in the vicinity of T_g (Fig. 2 A, curve 6). This implies that, at this temperature deformation of PMMA samples is accompanied by relaxation of low-temperature mode of the spectrum of segmental mobility during compression as a result of the appearance of segmental mobility within structural sublevels with lower packing densities. Hence, relaxation of ε_{res} takes place at temperatures close to T_g because of temperature-induced segmental mobility in the whole polymer sample.

For PMMA samples compressed at T_g, relaxation of ε_{res} involves two components: low-temperature component within glassy state and high-temperature component in the vicinity of T_g (Fig. 2 B, curve 2). The results discussed above allow one to conclude that the appearance of two-stage recovery of ε_{res} is likely to be related to an increase in T_g during compression. At the initial stage, deformation takes place at T_g (393 K). In this case, during compression relaxation of low-temperature mode of the spectrum of segmental mobility is observed, and accumulation of high-temperature component of relaxation takes place. As a result of loading, T_g of PMMA increases over 393 K, and, hence, further deformation proceeds at temperatures below T_g. Accumulation of low-temperature component of relaxation in plastically deformed samples therewith takes place.

The effect of pressure on glass transition temperature of polymers is widely discussed in literature (for example, in [11 - 14]). In most cases, with increasing pressure, a linear growth in T_g was observed.

To obtain a detailed information concerning the pressure dependence of T_g of PMMA, we studied the temperature dependence of compressibility factor β in the wide range of pressures (Fig. 1). Let us emphasize that, under compression in the press mold or under hydrostatic compression, physical and mechanical behavior is primarily controlled by the ratio between the loading time and relaxation time. Relaxation time is a function both of temperature and pressure. With increasing hydrostatic pressure, a significant increase in relaxation time is observed. From this standpoint, the experimental procedures for studing physical and mechanical behavior of polymers under the parallel action of pressure and temperature on loading or heating with a constant rate are not quite correct. In this case, the ratio between the loading time and relaxation time is different at each stage of test. To overcome these difficulties, in this work a step-by-step increase in pressure at fixed temperature was used. At each pressure, the samples were allowed to relax until the equilibrium state was attained.

Temperature interval corresponding to the jumplike change in compressibility factor β (Fig. 1) is associated with the glass transition of polymer at this pressure. Pressure dependence of T_g is shown in Fig. 3. This dependence is characterized by three well-pronounced regions. At pressures below 0.09 - 0.1 GPa, T_g is almost independent of the applied pressure. Within the pressure range from 0.09 - 0.1 to 0.38 - 0.4 GPa, with increasing temperature, the linear growth of T_g is observed. At pressures about 0.38 - 0.4 GPa, the pressure dependence of T_g shows an inflection point. The further increase in pressure leads to a slight increase in T_g.

Earlier [10], the dependence of T_g on the applied pressure was discussed when monolithization of powderlike PMMA was studied under the parallel action of pressure and temperature. In Fig. 4, curve 1 bounds the region of pressures and temperatures where monolithic PMMA samples are produced. These samples are monolithic up to temperatures corresponding to glass transition of PMMA. This reasoning allows one to conclude that the upper branch of curve 1 (Fig. 4) is the pressure dependence of T_g. Note that, in the case of monolithization of powderlike PMMA the pressure dependence of T_g (upper branch of the curve 1, Fig. 4) is well correlated with the pressure dependence of T_g obtained in this work for bulk polymer (curve 2).

Hence, for PMMA samples the pressure dependence of T_g is characterized by two inflection points at pressures 0.09 - 0.1 GPa and 0.38 - 0.4 GPa. Let us emphasize that for the compression of PMMA samples in the press mold at 293 K (Fig. 5 A, curve 1), the stress-strain diagram shows the inflection points at the same pressures. Such behavior of T_g with respect to the apllied pressure and compression stress-strain diagram may be explained in terms of the structural inhomogeneity of polymer glasses.

Note that, at pressures below 0.1 GPa stress-strain diagrams describing compression of PMMA samples in the mold (Fig. 5 A, curve 1) and uniaxial compression of PMMA samples (Fig. 5 A, curve 2) agree very closely. Let us emphasize that, the pressure about 0.1 GPa corresponds to yield point σ_y in the case of iniaxial compression of polymer. On the other hand, as it was shown in [10], at the pressure about 0.1 GPa, the plastic deformation in powderlike PMMA comes to a play (inflection in σ - Δh diagram, Fig. 5 B). It leads to monolithization of powderlike polymer due to diffusion of segments through the boundaries of particles of powder.

Similar profiles of the stress-strain diagrams corresponding to uniaxial compression of PMMA samples and compression in the press mold are likely to be associated with the fact that, in this range of pressures, the compression is controlled by the deformation of structural regions with lower packing densities. In the case of uniaxial compression, this assumption is supported by the direct structural studies of deformation of PMMA using positron annihilation lifetime spectroscopy [7 - 9]. This method allows one to identify structural sublevels with different packing densities in polymer glasses and to estimate concentration and sizes of free volume microregions within the above sublevels. As was shown, at initial portion of stress-strain diagram, the uniaxial compression of polymer samples is accompanied by the deformation of structural sublevels with lower packing densities via the decrease in the concentration of the microregions of free volume. Concentration of microregions of free volume in the more densely packed structural sublevels remains unchanged. This experimental evidence allows one to conclude that, at these pressures these structural regions are not involved in the deformation.

In this range of pressures the constant values of T_g correlates well with the results reported in [15 - 18], in which specific dual transitions were observed in the temperature region of glass transition. This fact was treated in terms of the existence of structural regions with different packing densities. Low-temperature transition is attributed to the appearance of segmental mobility in the loosely packed regions, whereas the high-temperature transition is provided by the segmental mobility in the densely packed domains. This high-temperature transition is associated with the macroscopic glass transition, when segmental mobility is realized in the whole polymer sample. When the densely packed structural regions are not involved in the deformation at pressures below 0.1 GPa, segmental mobility within these structural sublevels remains unchanged, and macroscopic T_g is constant.

When pressure increases above 0.1 GPa, the deformation behavior of polymer under compression in the press mold is quite different as compared with that under uniaxial compression (Fig. 5 A). The development of plastic deformation under uniaxial compression is accompanied by an increase in lateral sizes of the polymer samples at pressures about 0.1 GPa, that is, when σ_y is attained. In the case of compression in the press mold, lateral sizes of the sample remain unchanged. Plastic deformation of polymer is controlled by relatively uniform deformation of the densely packed structural sublevels. With increasing pressure, this process leads to densification of these structural regions, restriction of segmental mobility within these regions, and, finally, to the growth in T_g (Fig. 3).

As pressure increases to 0.4 GPa, the deformability of the densely packed structural regions is likely to be exhausted. Further increase in pressure is accompanied by a slight growth in strain (Fig. 5 A, curve 1) and T_g (Fig. 3).

CONCLUSION

Let us summarize the experimental evidence obtained.

1. At temperatures far below T_g, compression of bulk PMMA in the press mold is accompanied by the development of plastic deformation and accumulation of residual deformation. Relaxation of the residual deformation takes place at temperatures below T_g within glassy state of polymer;
2. The pressure dependence of T_g as well as the compression stress-strain diagram of PMMA show two well-pronounced inflection points at pressures of 0.09 - 0.1 and 0.38 - 0.4 GPa;
3. At pressures below 0.1 GPa, the stress-strain diagrams corresponding to uniaxial compression and compression in the press mold coincide.

These experimental evidence proves earlier speculations

concerning the structural pattern for plastic deformation of polymer glasses under uniaxial compression. These speculations are based on the ideas concerning the gradual involvement of various structural sublevels with different packing densities in deformation.

At pressures below 0.09 - 0.1 GPa plastic deformation is controlled by stress-induced segmental mobility within structural sublevels with lower packing densities. In this case, the stress-strain diagrams for uniaxial compression and compression in the press mold are quite similar , and T_g remains unchanged independently on applied pressure.

At pressures close to σ_y, that is, about 0.1 GPa, the densely packed structural sublevels are involved in plastic deformation. In the case of uniaxial compression, the development of plastic deformation is accompanied by the increase in lateral sizes of polymer samples. Under compression in the press mold the lateral sizes remain unchanged, and plastic deformation is controlled by a relatively uniform deformation of the densely packed structural regions. This leads to an increase in T_g because of the restriction of segmental mobility within these structural sublevels.

At pressures about 0.38 - 0.4 GPa, deformability of the densely packed structural sublevels is exhausted, and further increase in the applied pressure leads to a slight growth in strain and T_g.

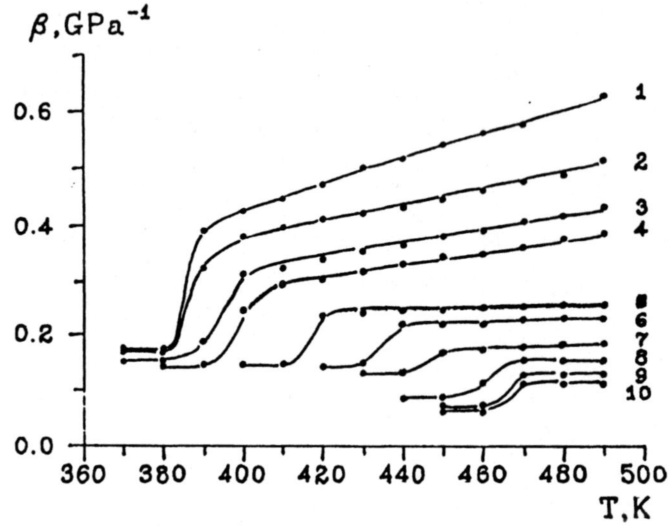

Fig. 1. Temperature dependences of compressibility factor β for PMMA at pressures: 0.03 (1), 0.06 (2), 0.09 (3), 0.12 (4), 0.21 (5), 0.3 (6), 0.39 (7), 0.48 (8), 0.6 (9), and 0.66 GPa (10).

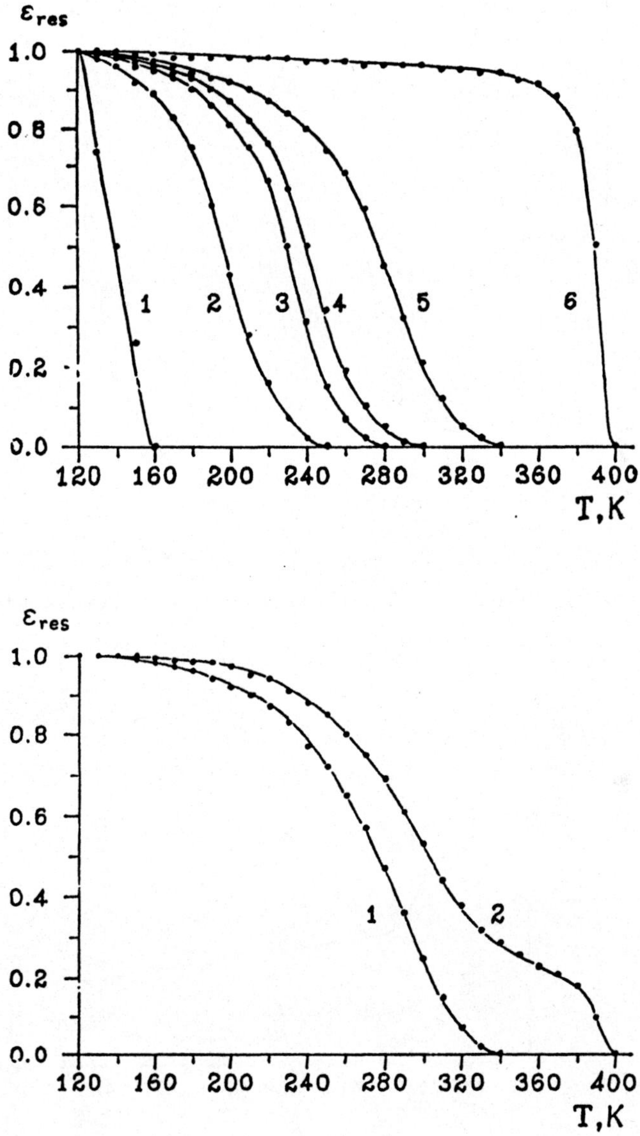

Fig. 2. Temperature dependences of residual deformation ε_{res} for PMMA samples compressed to 0.3 GPa at temperatures 173 (1), 233 (2), 273 (3), 293 (4), 353 (5), and 463 K (6) (A); and to 0.5 GPa at temperatures 353 (1) and 393 K (2) (B).

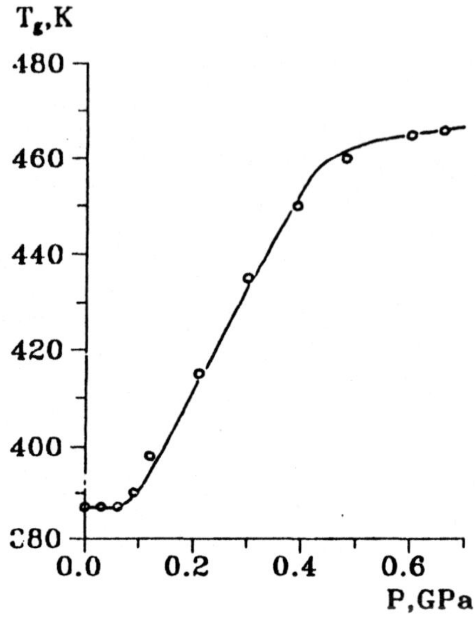

Fig. 3. Pressure dependence of glass transition temperature T_g of PMMA.

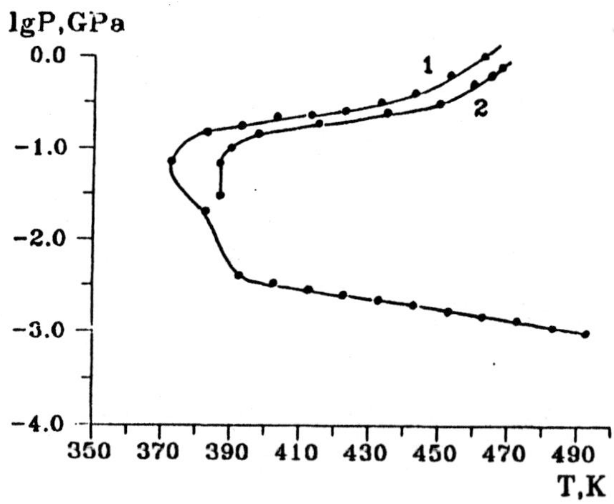

Fig. 4. "Pressure-temperature" diagram for preparation of monolithic PMMA samples from powderlike polymer (curve 1) and pressure dependence of glass transition temperature of bulk PMMA (curve 2).

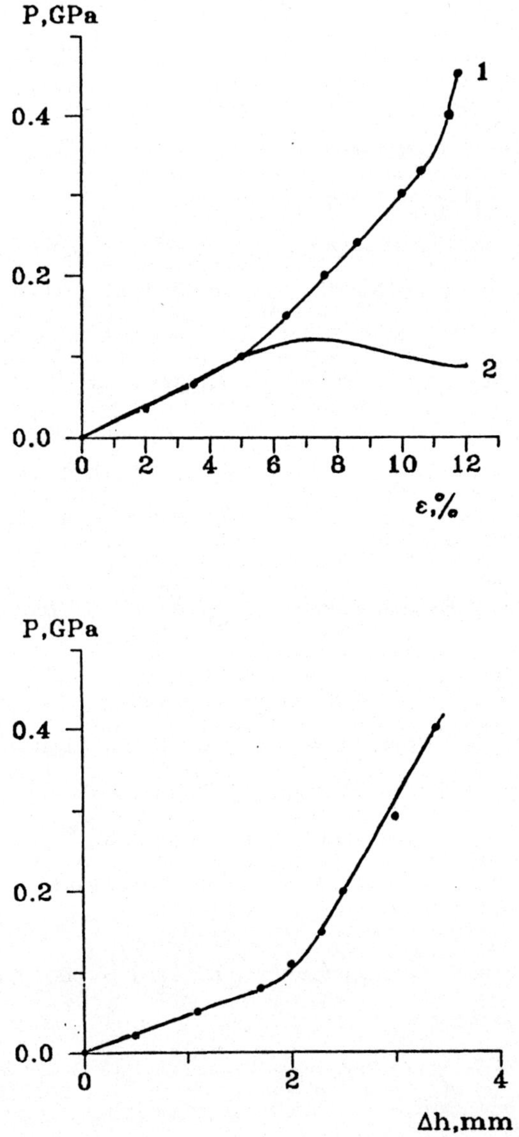

Fig. 5. Stress-strain diagrams for the compression of bulk PMMA samples in the press mold (1) and for the uniaxial compression of PMMA samples (2) (A), and for the compression of powderlike PMMA in the press mold (B).

REFERENCES

1. S.A. Arzhakov and V.A. Kabanov, Vysokomol. Soedin., B, 13, 318 (1971) (in Russian).

2. S.A. Arzhakov, N.F. Bakeev and V.A. Kabanov, Vysokomol. Soedin, A, 15, 1154 (1973) (in Russian).

3. E. Oleynik, Progr. Colloid & Polym. Sci., 80, 140 (1989).

4. E.F. Oleinik, O.B. Salamatina, S.N. Rudnev and S.V. Shenogin, Polym. Sci., 35, 1532 (1993).

5. M.S. Arzhakov and S.A. Arzhakov, Intern. J. Polymeric Mater., 29, 249 (1995).

6. M.S. Arzhakov and S.A. Arzhakov, in: Fizikokhimiya polimerov (Physics and Chemistry of Polymers), Tver', 2, 64 (1996) (in Russian).

7. I.B. Kevdina, M.S. Arzhakov and V.P. Shantarovich, Polymer Science, B, 37, 171 (1995).

8. M.S. Arzhakov, S.A. Arzhakov, I.B. Kevdina and V.P. Shantarovich, in: Fizikokhimiya polimerov (Physics and Chemistry of Polymers), Tver', 2, 59 (1996) (in Russian).

9. M.S. Arzhakov, S.A. Arzhakov, V.V. Gustov, I.B. Kevdina and V.P. Shantarovich, Intern. J. Polymeric Mater., (1996) (in press).

10. M.S. Arzhakov and S.A. Arzhakov, Intern. J. Polymeric Mater., (1996) (in press).

11. S.B. Ainbinder, K.I. Alkne, E.P. Tyunina, and M.G. Laka, Svoistva Polimerov pri Vysokikh Davleniyakh (Properties of Polymers at High Pressures), Moscow: Khimiya, 1973 (in Russian).

12. V.F. Skorodumov and Yu.K. Godovskii, Vysokomol. Soedin., 35, 214 (1993) (in Russian).

13. J. Havlicek, M. Ylalsty, and J. Hrouz, J. Macromol. Sci., Phys., 21, 425 (1982).

14. C. Fanggao, G.A. Saunders, E.F. Lambson, R.N. Hampton, G. Carini, G. Di Marco and M. Lanza, J. Polym. Sci., Polym. Phys., 34, 425 (1996).

15. V.I. Surovtsev, S.S. Pelishenko, N.M. Zelenskaya-Surovtseva, A.A. Savitskaya and V.A. Degtyarev, Vysokomol. Soedin., A, 31, 339 (1989) (in Russian).

16. V.N. Belousov, B.Kh. Kotsev, and A.K. Mikitaev, Dokl. Akad. Nauk SSSR, 280, 1140 (1985) (in Russian).

17. I.I. Perepechko and O.V. Startsev, Akusticheskii Zhurnal, 22, 749 (1976) (in Russian).

18. I.I. Perepechko, Akusticheskie metody issledovaniya polimerov (Acoustic Studies of Polymers), Moscow: Khimiya, 1973 (in Russian).

Physical and Mechanical Behavior of Polymer Glasses III. Copolymers of Methyl Methacrylate

M.S. Arzhakov, S.A. Arzhakov, M.B. Lachinov, E.V. Nechaeva and D.V. Trachenko

Polymer Department, Faculty of Chemistry, Lomonosov Moscow State University, Vorob'evy Gory, Moscow, 119899 Russia

Physical and mechanical behavior of copolymers of methyl methacrylate and PMMA was studied. The contribution from low-temperature component to total temperature-induced relaxation of plastic deformation was shown to correlate well with the mechanical characteristics of polymer samples. This correlation was discussed in terms of the structural inhomogeneity of polymer glasses: coexistence of structural sublevels with characteristic values of relaxation time and energy of activation of segmental mobility. From this standpoint, plastic deformation and, in general, physical and mechanical behavior of polymer glasses is likely to be controlled by a gradual involvement of segmental mobility within above structural sublevels in deformation and relaxation processes.

Key words: glassy polymers, copolymers of methyl methacrylate, plastic deformation, relaxation

INTRODUCTION

A two-stage character of temperature-induced relaxation of plastic deformation of polymer glasses, that is, existence of low-temperature and high-temperature components of relaxation, is considered as a manifestation of structural rearrangements which are responsible for and accompany plastic deformation [1 - 4]. In this case, physical and mechanical properties of glassy polymers are likely to be controlled by the coexistence of the above two components of temperature-induced relaxation.

The earlier studies concerning physical and mechanical behavior of PMMA samples [3, 4] allowed us to reveal a certain correlation between the mechanical properties of polymer and the contributions from low-temperature and high-temperature components to total temperature-induced relaxation of plastic deformation. From this standpoint, the ratio between these components of relaxation is treated as a new physical parameter, which was advanced to describe plastic deformation in glassy polymers. This reasoning is based on the speculations concerning the structural pattern of plastic deformation of polymer glasses, that is, a gradual involvement of segmental mobility within various structural sublevels in deformation and relaxation processes.

To gain a deeper insight into the nature of low-temperature and high-temperature components of relaxation and their role in development of polymer deformation, we studied physical and mechanical behavior of copolymers of methyl methacrylate with methacrylic monomers.

EXPERIMENTAL

In this work, random copolymers of methyl methacrylate (MMA) with butyl methacrylate (BMA), octyl methacrylate (OMA), and lauryl methacrylate (LMA) were used.

Prior to copolymerization, the monomers were distilled under vacuum in an nitrogen flow. The initiator of copolymerization (lauroyl peroxide) was purified by recrystallization from ethanol.

The monomer feed composition was the following: for MMA/BMA - 95/5, 90/10, 80/20, 70/30, and 50/50; for MMA/OMA - 95/5, 90/10, and 80/20; for MMA/LMA - 95/5, 90/10, and 85/15. The monomer mixtures with initiator were deoxygenated by repeated freezing-defreezing at a pressure of 10^{-2} mmHg. For all mixtures, the content of lauroyl peroxide was $5x10^{-3}$ mol/l.

Copolymerization was carried out in the sealed glass tubes with a diameter of 10 mm and a height of 100 mm at 333 K under vacuum. To achieve the complete conversion, at the final stage of copolymerization reactions temperature was increased up to temperatures, which exceeded glass transition temperatures of the corresponding copolymers by 10 - 15 K.

Glass transition temperatures of the copolymers studied were estimated by differential scanning calorimetry using "Mettler TA-4000". Heating rate was 20 K/min. Glass transition temperatures of copolymer samples are summarized in Table 1.

For mechanical tests, the cylindrical copolymer samples with a height of 5 mm and a diameter of 5 mm were used. Prior to testing, the samples were annealed at temperatures above glass transition temperature by 10 - 15 K. Then the samples were slowly cooled down to room temperature. The height of the as-prepared samples was denoted as h_0.

The test samples were uniaxially compressed at 293 K with a strain rates 0.1, 1, 10, and 100 %/min using an UTS-10 tensile machine (Germany). Strain was 20%. Yield stress σ_y was estimated with an accuracy of ± 5%. The height of stressed sample was denoted as h_{def}. Then, the samples were unloaded at 293 K with the same rates.

In the deformed polymer samples, temperature-induced relaxation of residual deformation was studied by measuring the changes in the height of the test samples on their heating in the temperature range from 293 K to T_g. The free-standing samples were allowed to relax at certain temperatures within the above temperature interval until the equilibrium height was attained. The equilibrium height of samples at a given temperature was denoted as h_T. Residual deformation at a given temperature was estimated as $\varepsilon_{res} = (h_0-h_T)/(h_0-h_{def})$ with an accuracy of ± 2%.

RESULTS AND DISCUSSION

Figure 1 shows stress-strain diagrams describing uniaxial compression of the copolymer samples. As follows from Fig. 1, with increasing the concentration of BMA, OMA, and LMA in their copolymers with MMA, a well-pronounced decrease in mechanical characteristics of polymer, in particular, yield stress σ_y, is observed. The values of σ_y are summarized in Table 1.

For plastically deformed copolymer samples, the temperature

dependences of residual deformation ε_{res} are shown in Fig. 2. Note, that temperature-induced relaxation of residual deformation takes place in two temperature regions and involves two components: low-temperature component ε_1 and high-temperature component ε_2. These components of relaxation are shown in Fig. 1A for the temperature dependence of ε_{res} of plastically deformed PMMA (curve 1).

A complete relaxation is observed in a narrow temperature range in the vicinity of glass transition temperature T_g of copolymers via high-temperature component ε_2. However, relaxation of a marked fraction of plastic deformation is observed in the wide temperature range at temperatures well below T_g via low-temperature component ε_1.

Earlier [3 - 5], two-stage character of temperature-induced relaxation of plastic deformation of polymer glasses was explained by structural inhomogeneity of polymer glasses, that is, coexistence of structural sublevels with different packing densities and degrees of ordering. Each of these sublevels is characterized by its own relaxation time and activation energy of segmental mobility. Hence, at each temperature below glass transition temperature local segmental mobility in certain structural sublevels is allowed. With increasing temperature, structural sublevels with higher packing densities become involved in segmental mobility, and finally, at T_g, segmental mobility is observed in the whole polymer sample. In other words, the existence of a set of structural sublevels with different packing densities is responsible for the spectrum of segmental mobility in glassy polymer.

Let us emphasize, that a possible segmental mobility in the local structural sublevels at temperatures below T_g is widely discussed in literature (for example, in [6 - 10]).

The low-temperature component ε_1 of relaxation of residual deformation is associated with the low-temperature mode of the spectrum of segmental mobility due to the temperature-induced local segmental mobility, which is observed within structural sublevels with lower packing densities at temperatures well below glass transition temperature. A complete relaxation of plastic deformation is accomplished at glass transition and is provided by high-temperature component ε_2, when segmental mobility is observed in a whole polymer sample. Hence, molecular mechanism of low-temperature component of relaxation of residual deformation seems to be similar to that of high-temperature component and involves the transition of the excited chain conformations to initial equilibrium state via segmental mobility.

The effect of strain on the ratio between ε_1 and ε_2 was studied in [3]. As shown in [3], in the case of glassy polymers with strains below or equal to yield strain ε_y, relaxation of residual deformation is controlled only by low-temperature component ε_1, that is, by temperature-induced segmental mobility within the structural sublevels with lower packing densities. This evidence implies, that polymer deformation to strains below yield strain is associated with stress-induced local segmental mobility within these structural sublevels. With increasing strain above yield point, at constant ε_1 appearance and growth of high-temperature component of relaxation ε_2 is observed. Relaxation of high-temperature component ε_2 is controlled by temperature-induced segmental mobility in the whole polymer sample. This reasoning suggests that, at strains above yield strain, all segments in polymer sample are involved in plastic deformation. Hence, two components of relaxation of residual deformation allow one to advance some speculations concerning the structural pattern of plastic deformation. According to these speculations, for polymer glasses plastic deformation is controlled by a gradual involvement of the structural sublevels with different packing densities in deformation and relaxation processes. In other words, plastic deformation of polymer glasses may be described by the contributions from low-temperature and high-temperature components to relaxation of residual deformation. From this standpoint, finding correlation between the contributions from these components to relaxation of residual deformation and polymer strength properties is necessary to describe physical and mechanical behavior of polymer glasses.

With respect to polymer strength properties, consideration of the contribution from low-temperature component to total relaxation of residual deformation $\varepsilon_1/(\varepsilon_1+\varepsilon_2)$ seems to be of most interest. This parameter concerns the development of polymer deformation at initial portion of stress-strain diagram at strains below or equal to ε_y, which is associated with such strength properties as elastic modulus E_0 and σ_y. Earlier [3], the parallel decrease in E_0, σ_y, and $\varepsilon_1/(\varepsilon_1+\varepsilon_2)$ was observed as a result of increasing deformation temperature T_{def} of PMMA as well as plasticization of PMMA with dibutyl phtalate (DBPh). For PMMA samples deformed at different T_{def} and PMMA plasticized with DBPh, the values of σ_y and $\varepsilon_1/(\varepsilon_1+\varepsilon_2)$ are summarized in Table 1.

As evidenced by Table 1, with increasing concentration of BMA, OMA, and LMA in copolymers with MMA, decrease in both σ_y and $\varepsilon_1/(\varepsilon_1+\varepsilon_2)$ is observed. Let us consider the correlation between the above parameters for all polymer samples studied.

For copolymers of MMA as well as for PMMA deformed at different T_{def} and plasticized PMMA, Fig. 3 shows the dependence of σ_y on $\varepsilon_1/(\varepsilon_1+\varepsilon_2)$. As follows from Fig. 3, all experimental data fit one curve. Hence, there is a good correlation between σ_y and $\varepsilon_1/(\varepsilon_1+\varepsilon_2)$ and this correlation is independent of any variations in these parameters: either by varying T_{def} of PMMA, or plasticization of PMMA, or copolymerization of MMA with BMA, OMA, and LMA. This fact is likely to be associated with the difference between T_g and T_{def}, which is defined as $\Delta T_{def} = T_g - T_{def}$.

Experimental evidence allows one to estimate both σ_y and $\varepsilon_1/(\varepsilon_1+\varepsilon_2)$ at a certain ΔT_{def} (Table 2). At a given ΔT_{def}, σ_y and $\varepsilon_1/(\varepsilon_1+\varepsilon_2)$ remain unchanged and are independent of any variations in ΔT_{def}: by increasing T_{def} at a fixed T_g or by decreasing T_g at a fixed T_{def}. Depression in T_g was attained by either plasticization of PMMA or by copolymerization of MMA with BMA, OMA, and LMA. Note, that such behavior is likely to be associated with the dynamics of segmental mobility in polymer glasses.

As was mentioned earlier, low-temperature relaxation of residual deformation at temperatures well below T_g proceeds via low-temperature mode of the spectrum of segmental mobility within structural sublevels with lower packing densities. Obviously, a certain fraction of low-temperature mode of segmental mobility is able to relax during polymer loading. From this viewpoint, low-temperature component ε_1 is likely to be associated with stored plastic deformation, which remained unrelaxed during deformation. With decreasing ΔT_{def}, local segmental mobility in certain structural regions is enhanced. As a result, the higher fraction of low-temperature mode of the whole spectrum of segmental mobility is able to relax during deformation, and decrease in both $\varepsilon_1/(\varepsilon_1+\varepsilon_2)$ and σ_y is observed.

The dynamics of segmental mobility seems to be controlled not only by ΔT_{def} but also by strain rate [11, 12]. One could expect, that with increasing strain rate, the fraction of low-temperature mode of segmental mobility, which is able to relax during deformation, decreases. As a result, both σ_y and $\varepsilon_1/(\varepsilon_1+\varepsilon_2)$ should increase. To verify this idea, copolymers of MMA with BMA, OMA, and IMA were tested under the compression with different strain rates.

For copolymer samples, with increasing strain rate from 0.1 to 100 %/min, increase in both σ_y and $\varepsilon_1/(\varepsilon_1+\varepsilon_2)$ is observed. Figure 4 shows the dependence of σ_y on $\varepsilon_1/(\varepsilon_1+\varepsilon_2)$, when these parameters were varied by changing strain rate. In this case, a well-pronounced correlation between σ_y and $\varepsilon_1/(\varepsilon_1+\varepsilon_2)$ is also observed.

CONCLUSION

Low-temperature and high-temperature components of temperature-induced relaxation of residual plastic deformation are related to structural inhomogeneity of polymer glasses: coexistence of structural sublevels with characteristic relaxation time and energy of activation of segmental mobility. In other words, the spectrum of segmental mobility in glassy polymer is controlled by structural inhomogeneity in polymer glasses.

The ratio between low-temperature and high-temperature components of relaxation of residual deformation provides an information concerning the dynamics of segmental mobility within certain structural sublevels during loading. For copolymers of MMA with BMA, OMA, and LMA and for PMMA samples, there is an evident correlation between the above ratio and yield stress, and this correlation is independent of any variations in these parameters: either by varying $\Delta T_{def} = T_g - T_{def}$ or by varying strain rate. This experimental evidence allows one to conclude that plastic deformation and, in general, physical and mechanical behavior of polymer glasses is likely to be controlled by segmental mobility within different structural sublevels and their contributions to deformation and relaxation.

Table 1.

Physical and mechanical parameters of PMMA and various copolymers of methyl methacrylate

Sample	T_g,K	T_{def},K	σ_y,GPa	$\varepsilon_1/(\varepsilon_1+\varepsilon_2)$	ΔT_{def},K
PMMA	393	293	0.124	0.88	100
		320	0.082	0.78	73
		333	0.070	0.72	60
		353	0.060	0.61	40
		383	0.050	0.50	10
PMMA + 5 wt% of DBPh	383	293	0.102	0.85	90
PMMA + 10 wt% of DBPh	373	-	0.086	0.78	80
PMMA + 20 wt% of DBPh	343	-	0.065	0.68	50
MMA/BMA					
95/5	383	-	0.115	0.88	90
90/10	373	-	0.110	0.88	80
80/20	363	-	0.094	0.85	70
70/30	353	-	0.083	0.82	60
50/50	333	-	0.060	0.70	40
MMA/OMA					
95/5	378	-	0.096	0.83	85
90/10	363	-	0.075	0.75	70
80/20	343	-	0.055	0.63	50
MMA/LMA					
95/5	368	-	0.090	0.78	75
90/10	353	-	0.060	0.68	60
85/15	338	-	0.050	0.57	45

Table 2

Yield stress σ_y and contribution from low-temperature component to relaxation of residual deformation $\varepsilon_1/(\varepsilon_1+\varepsilon_2)$ of PMMA and various copolymers of methyl methacrylate at a given ΔT_{def}.

ΔT_{def}, K	σ_y, GPa	$\varepsilon_1/(\varepsilon_1+\varepsilon_2)$
80 - 90	0.105 ± 0.010	0.86 ± 0.02
70 - 80	0.090 ± 0.005	0.81 ± 0.04
60 - 70	0.065 ± 0.005	0.70 ± 0.02
40 - 50	0.055 ± 0.005	0.60 ± 0.03

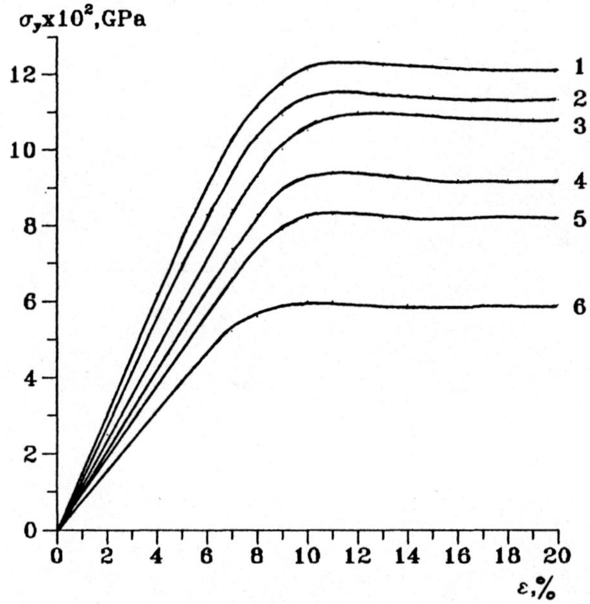
σy x10², GPa
12
10
8
6
4
2
0
0 2 4 6 8 10 12 14 16 18 20
ε,%
1
2
3
4
5
6

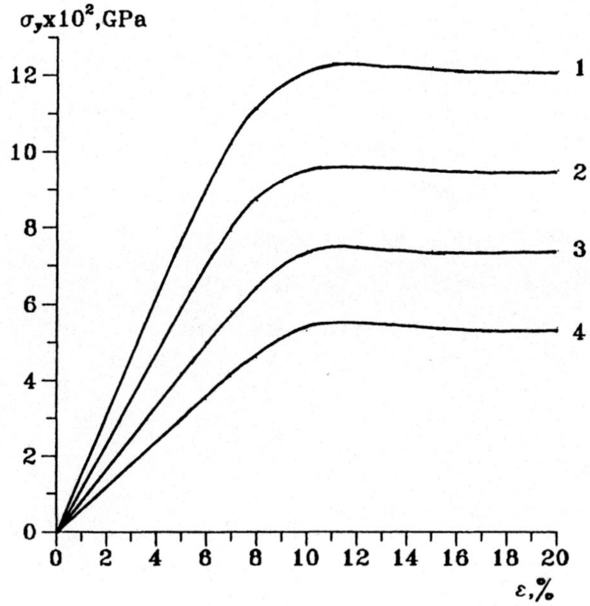
σy x10², GPa
12
10
8
6
4
2
0
0 2 4 6 8 10 12 14 16 18 20
ε,%
1
2
3
4

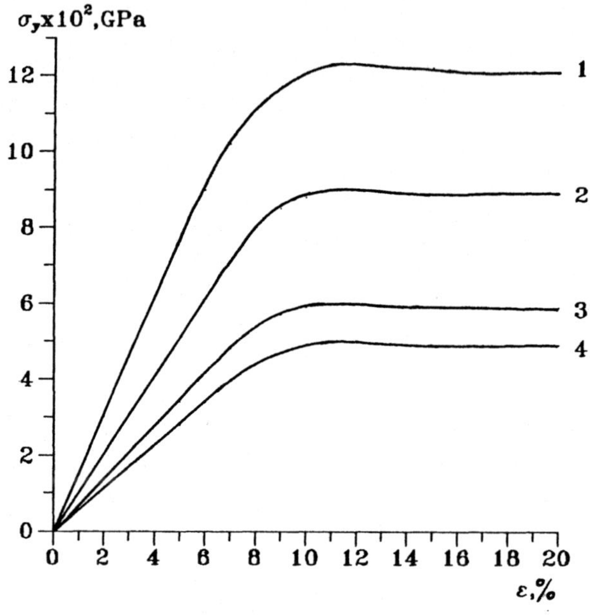

Fig. 1. Stress-strain diagrams of PMMA (curves 1) and various copolymers of MMA. Monomer feed composition: MMA/BMA - 95/5 (2), 90/10 (3), 80/20 (4), 70/30 (5), and 50/50 (6) (A); MMA/OMA - 95/5 (2), 90/10 (3), and 80/20 (4) (B); MMA/LMA - 95/5 (2), 90/10 (3), and 85/15 (4) (C). Uniaxial compression at 293 K. Strain rate 1 %/min.

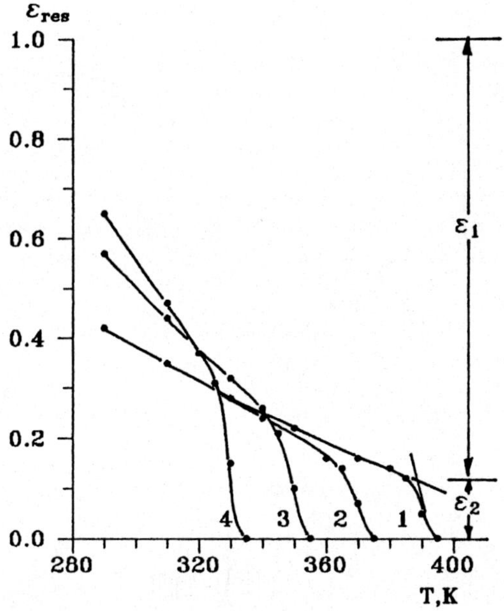
ε_{res}
1.0
0.8
0.6
0.4
0.2
0.0
280
320
360
400
T,K
ε_1
ε_2
4
3
2
1

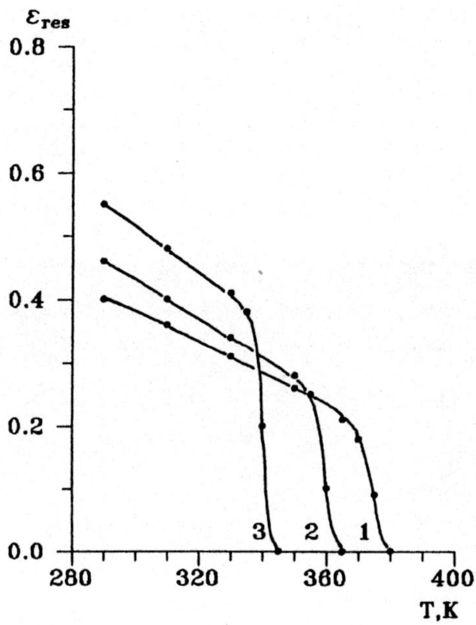
ε_{res}
0.8
0.6
0.4
0.2
0.0
280
320
360
400
T,K
3
2
1

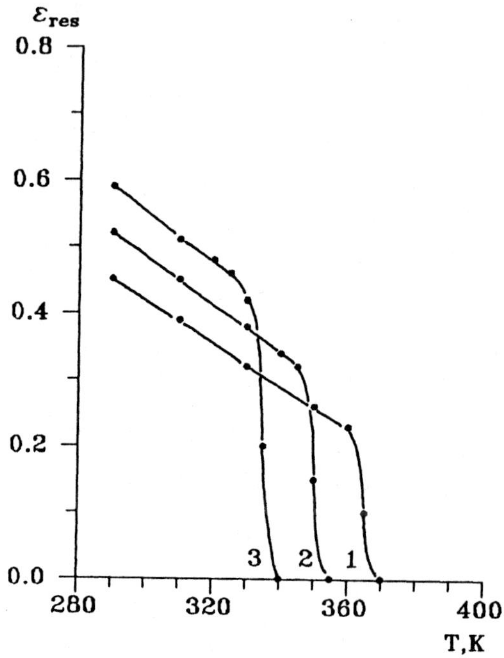

Fig. 2. Temperature dependences of residual deformation ε_{res} of various copolymers of MMA. Monomer feed composition: MMA/BMA - 100/0 (1), 90/10 (2), 70/30 (3), and 50/50 (4) (A); MMA/OMA - 95/5 (1), 90/10 (2), and 80/20 (3) (B); MMA/LMA - 95/5 (1), 90/10 (2), and 85/15 (3) (C).

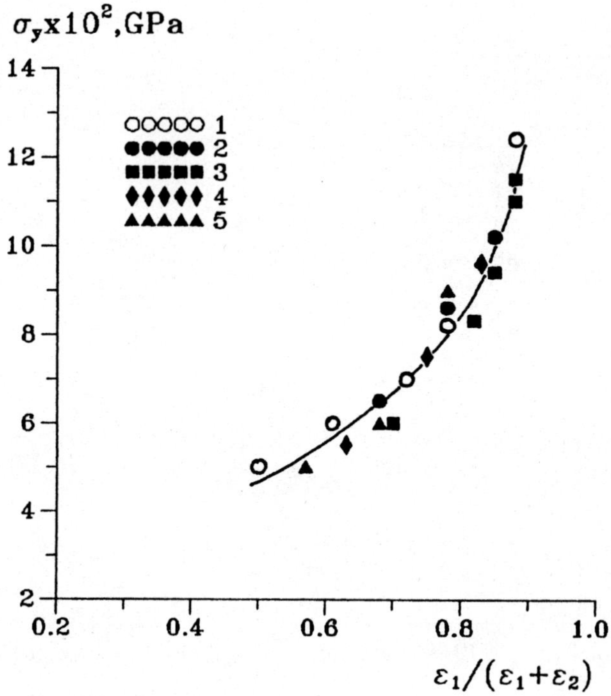

Fig. 3. Yield stress σ_y versus contribution from low-temperature component to relaxation of residual deformation $\varepsilon_1/(\varepsilon_1+\varepsilon_2)$ for PMMA compressed at different temperatures (1), plasticized PMMA (2), and copolymers of MMA with BMA (3), OMA (4), and LMA (5).

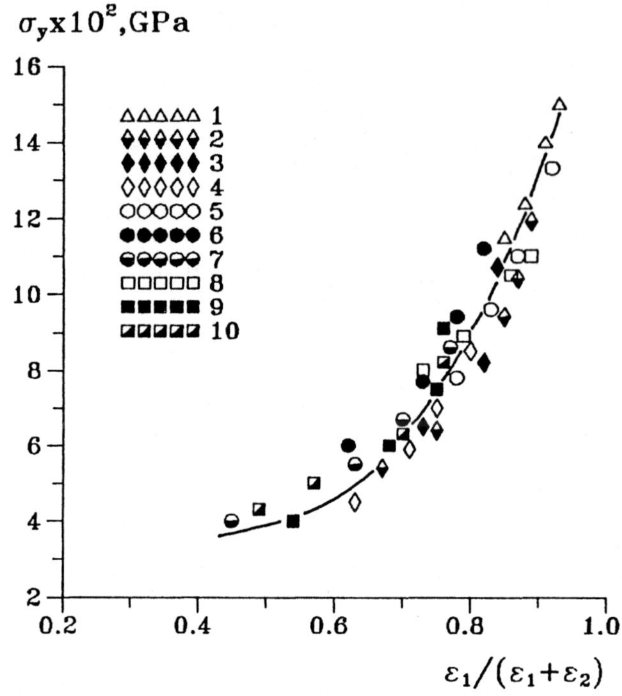

Fig. 4. Yield stress σ_y versus contribution from low-temperature component to relaxation of residual deformation $\varepsilon_1/(\varepsilon_1+\varepsilon_2)$ for PMMA (1) and various copolymers of MMA. Monomer feed composition: MMA/BMA - 80/20 (2), 70/30 (3), and 50/50 (4); MMA/OMA - 95/5 (5), 90/10 (6), and 80/20 (7); MMA/LMA - 95/5 (8), 90/10 (9), and 85/15 (10). Comments are given in text.

REFERENCES

1. E. Oleynik, Progr. Colloid & Polym. Sci., 80, 140 (1989).

2. E.F. Oleinik, O.B. Salamatina, S.N. Rudnev and S.V. Shenogin, Polym. Sci., 35, 1532 (1993).

3. M.S. Arzhakov and S.A. Arzhakov, Intern. J. Polymeric Mater., 29, 249 (1995).

4. M.S. Arzhakov and S.A. Arzhakov, in: Fizikokhimiya polimerov (Physics and Chemistry of Polymers), Tver', 2, 64 (1996) (in Russian).

5. M.S. Arzhakov, S.A. Arzhakov, I.B. Kevdina and V.P. Shantarovich, in: Fizikokhimiya polimerov (Physics and Chemistry of Polymers), Tver', 2, 59 (1996) (in Russian).

6. V.A. Bershtein and V.M. Egorov, Differentsial'naya skaniruyushchaya kalorimetriya v fizikokhimii polimerov (Differential Scanning Calorimetry in Physical Chemistry of Polymers), Leningrad: Khimiya, 1990 (in Russian).

7. V.I. Surovtsev, S.S. Pelishenko, N.M. Zelenskaya-Surovtseva, A.A. Savitskaya and V.A. Degtyarev, Vysokomol. Soedin., A, 31, 339 (1989) (in Russian).

8. V.N. Belousov, B.Kh. Kotsev and A.K. Mikitaev, Dokl. Akad. Nauk SSSR, 280, 1140 (1985) (in Russian).

9. I.I. Perepechko, Akusticheskie metody issledovaniya polimerov (Acoustic Studies in Polymers), Moscow: Khimiya, 1973 (in Russian).

10. G.V. Kozlov, V.N. Shogenov and A.K. Mikitaev, Dokl. Akad. Nauk SSSR, 298, 142 (1988) (in Russian).

11. The Physics of Glassy Polymers, R.N. Haward, Ed., New York, Toronto: Wiley, 1973.

12. L.E. Nielsen, Mechanical Properties of Polymers and Composites, New York: Dekker, 1974.

Physical and Mechanical Behavior of Polymer Glasses IV. Cross-Linked Poly(Methyl Methacrylate)

M.S. Arzhakov and S.A. Arzhakov

Polymer Department, Faculty of Chemistry, Lomonosov Moscow State University, Vorob'evy Gory, Moscow, 119899 Russia

Temperature-induced relaxation of plastic deformation in cross-linked PMMA samples was studied. For cross-linked PMMA with low concentration of cross-linking agent (less, than 5.2 mol.%), relaxation curves involve low-temperature component at temperatures well below glass transition temperature and high-temperature component at temperatures close to glass transition. Increasing concentration of cross-linked agents in polymer samples to 11.2 mol.% is accompanied with degeneration of high-temperature component of relaxation, and complete relaxation of plastic deformation proceeds via low-temperature component. The effect of cross-linking on temperature-induced relaxation of plastic deformation was discussed in terms of restriction of segmental mobility within certain structural regions in polymer sample.

Key words: PMMA, cross-linking, plastic deformation, relaxation

INTRODUCTION

Temperature-induced relaxation of plastic deformation of polymer glasses is known to involve low-temperature component at temperatures well below glass transition temperature and high-temperature component in the vicinity of glass transition [1 - 4]. Earlier [5, 6], two-stage character of relaxation was discussed in terms of structural inhomogeneity of glassy polymers, that is, the existence of structural sublevels with different packing densities and degrees of ordering. The existence of a set of the above structural sublevels is responsible for the gradual involvement of segmental mobility in temperature-induced relaxation of plastically deformed polymer sample.

Decreasing contribution from low-temperature component to total relaxation of residual deformation was observed as a result of increasing deformation temperature of PMMA, plasticization of PMMA, and copolymerization of methyl methacrylate with methacrylic monomers [5, 7]. This experimental evidence was attributed to enhancement of local segmental mobility in certain structural sublevels. From this standpoint, the ratio between low-temperature and high-temperature components of temperature-induced relaxation of plastic deformation provides an important information concerning structural pattern of development of segmental mobility in glassy polymer. Generally speaking, this ratio may be treated as a physical parameter describing plastic deformation of polymer glasses, and their physical and mechanical behavior [5, 6].

In the present work, temperature-induced relaxation of plastic deformation of cross-linked PMMA was studied to gain a deeper insight in physical and mechanical behavior of glassy polymers.

EXPERIMENTAL

The copolymers of methyl methacrylate (MMA) with bifunctional monomers ethylene glycol dimethacrylate (EGDM) and triethylene glycol dimethacrylate (TGDM) were used as cross-linked PMMA. The content of bifunctional monomer in polymerization system was 2.5, 5.2, and 11.2 mol%. The development of network took place immediately during bulk polymerization.

Glass transition temperatures of copolymers studied were estimated by differential scanning calorimetry using a "Mettler TA-4000". Heating rate was 20 K/min.

The test samples were cylinders with a height of 30 mm and a diameter of 15 mm. Prior to testing, polymer samples were annealed at temperatures, which exceeded glass transition temperature by 10 - 15 K. Then, they were cooled down to room temperature. The height of as-prepared samples was denoted as h_o.

The samples of initial PMMA and PMMA, cross-linked with EGDM and TGDM, were uniaxially compressed at 293 K with strain rate of 0.1 mm/min using an UTS-100 tensile machine (Germany). After that, they were unloaded at this temperature with the same rate. The height of deformed samples was denoted as h_{def}.

In the deformed polymer samples, temperature-induced relaxation of residual deformation was studied by measuring the changes in their height on the heating from 293 K at a heating rate of 1 K/min. The height of the samples at a given temperature was denoted as h_T. Residual deformation at a given temperature was estimated as $\varepsilon_{res} = (h_o - h_T)/(h_o - h_{def})$ with an accuracy of ± 2%.

RESULTS AND DISCUSSION

For initial PMMA and PMMA, cross-linked with EGDM and TGDM, Fig. 1 shows stress-strain diagrams, corresponding to uniaxial compression of polymer samples. Note, that all samples show well-pronounced plastic deformation and storing residual deformation after their unloading. Let us emphasize, that for cross-linked samples their elastic modulus and yield stress seem to be independent of the concentration of cross-linking agent (EGDM and TGDM), and are equal to those for initial PMMA.

For initial PMMA and cross-linked PMMA, Figs. 2 and 3 show temperature dependences of residual deformation ε_{res}. For the lack of space, relaxation curves are presented only for polymer samples with 20 and 30% strains. As follows from Figs. 2 and 3, for PMMA samples (curves 1), temperature-induced relaxation of plastic deformation takes place in two well-defined temperature regions below glass transition temperature T_g and close to T_g via low-temperature component ε_1 and high-temperature component ε_2, respectively. These components of relaxation are shown in Fig. 2 A.

Earlier [5, 6], two-stage character of temperature-induced relaxation of residual deformation of polymer glasses was interpreted in terms of their structural inhomogeneity, that is, coexistence of structural sublevels with different packing densities and degrees of ordering. Each of these sublevels is characterized by its own relaxation time and activation energy of segmental mobility. Hence, at each temperature below glass transition temperature local segmental mobility in certain structural sublevels is allowed. With increasing temperature, structural sublevels with higher packing densities become involved in segmental mobility, and, finally, at glass transition temperature segmental mobility is observed in the whole polymer sample. In other words, the existence of a set of structural sublevels with different packing densities is responsible for the spectrum of segmental mobility in polymer glasses.

Low-temperature relaxation of residual deformation is controlled by low-temperature fraction of the spectrum of segmental mobility because of temperature-induced segmental mobility within structural sublevels with lower packing densities at temperatures well below T_g. Evidently, a certain part of low-temperature fraction of the spectrum of segmental mobility is able to relax during polymer loading. From this standpoint, low-temperature component ε_1 of relaxation is likely to be associated with stored plastic deformation, which remained unrelaxed during loading.

High-temperature fraction of the spectrum of segmental mobility is responsible for high-temperature component ε_2 of relaxation. This component is accumulated in polymer during loading, and is able to relax only at temperatures close to T_g via temperature-induced segmental mobility within structural sublevels with higher packing densities.

Hence, the ratio between low-temperature and high-temperature components of relaxation provides an important information concerning structural pattern of the development of temperature-induced segmental mobility in plastically deformed glassy polymer.

The effect of strain on the ratio between low-temperature and high-temperature components of relaxation [5, 6] in combination with direct structural studies in deformation using positron annihilation lifetime spectroscopy [8, 9] allowed us to advance the structural pattern of plastic deformation. Polymer deformation to strains below or equal to yield strain is associated with stress-induced local segmental mobility within structural sublevels with lower packing densities. With increasing strain above yield point, segmental mobility within structural sublevels with higher packing densities comes to a play. In other words, for polymer glasses plastic deformation is controlled by gradual involvement of segmental mobility within structural sublevels characterized by different packing densities in deformation and relaxation processes.

Earlier [5, 7], increasing deformation temperature, plasticization of PMMA, or copolymerization of methyl methacrylate with methacrylic monomers was shown to be accompanied with decrease in contribution from low-temperature component ε_1 to total temperature-induced relaxation of residual deformation. This fact is likely to be associated with enhancement in local segmental mobility in certain structural sublevels. As a result, the higher fraction of low-temperature mode of the whole spectrum of segmental mobility is able to relax during deformation, and decrease in low-temperature component ε_1 is observed.

This experimental evidence allows one to suggest that restriction of segmental mobility in polymer should result in the growth in contribution from low-temperature component ε_1 to relaxation. Restriction of segmental mobility in polymer sample may be achieved by cross-linking of polymers [10]. For PMMA samples, cross-linked with EGDM and TGDM, restriction of segmental mobility is evidenced by increasing T_g with increase in concentration of cross-linking agents in polymer samples. To verify the effect of cross-linking on the ratio between low-temperature and high-temperature components of relaxation, temperature-induced relaxation of cross-linked PMMA samples was studied.

For PMMA samples, cross-linked with 2.5 and 5.2 mol% of cross-linking agents (Figs. 2 and 3, curves 2 and 3), temperature dependences of residual deformation also are characterized by low-temperature and high-temperature components. A well-pronounced two-stage character of temperature-induced relaxation of residual deformation allows one to suggest that, for cross-linked samples, relaxation is controlled by temperature-induced segmental mobility within structural sublevels with different packing densities as for the initial PMMA. In other words, cross-linked PMMA samples are likely to be characterized by structural inhomogeneity. Note, that, at low concentration of cross-linking agents, morphology of cross-linked polymer is known to be similar to that for linear polymer [10]. This experimental evidence allows one to suggests, that, in these cases, cross-links are localized in both loosely and densely packed structural sublevels.

Cross-linking of structural sublevels with lower packing densities is responsible for the increase in low-temperature component ε_1 of relaxation because of restriction of segmental mobility within these structural regions.

Cross-linking of structural sublevels with higher packing densities is evidenced by stress-strain diagrams of cross-linked polymer samples (Fig. 1). As was mentioned above, post-yield behavior of polymer glasses is controlled by stress-induced segmental mobility in structural sublevels with higher packing densities. As follows from Fig. 1, for cross-linked PMMA samples, post-yield deformation proceeds at higher stress as compared with initial polymer. This fact may be attributed to the restriction of segmental mobility within these structural regions because of cross-linking.

With increasing concentration of EGDM and TGDM to 11.2 mol.% (Figs. 2 and 3, curves 4), high-temperature component ε_2 is degenerated, and temperature-induced relaxation of residual deformation proceeds via low-temperature component ε_1. Complete relaxation is observed at temperatures well below T_g within glassy state.

Such relaxation behavior may be explained from two standpoints. At first, increasing concentration of cross-linking agent in PMMA is likely to be responsible for the further cross-linking of structural sublevels with lower packing densities. As a result, a noticeable restriction of segmental mobility within these structural sublevels is observed, and total temperature-induced relaxation of residual deformation is primarily controlled by low-temperature component ε_1. Secondly, increasing concentration of cross-linking agent in polymerization system is known [10] to present a severe difficulties for formation of morphological structures with high degree of ordering. Finally, the degeneration of these structures is observed. In this case, one could expect one-stage recovery of residual deformation.

Hence, cross-linking of PMMA is accompanied by restriction of segmental mobility within certain structural sublevels in polymer samples. As a result, the lower fraction of low-temperature mode of the whole spectrum of segmental mobility is able to relax during deformation, and increase in contribution from low-temperature component ε_1 to total temperature-induced relaxation of residual deformation is observed. Obviously, restriction of segmental mobility should be controlled by the length of the molecule of cross-linking agent.

Figure 4 shows the temperature dependences of residual deformation ε_{res} for cross-linked PMMA samples with the similar concentration of EGDM and TGDM. When passing from EGDM to TGDM (curves 2 and 3, respectively), increasing the length of cross-links results in decrease in low-temperature component ε_1. Note, that, for the samples cross-linked with EGDM, post-yield deformation proceeds at higher stress as compared with the samples cross-linked with TGDM (Fig. 1). This behavior may be attributed to the fact, that the higher restriction of segmental mobility is observed for the samples cross-linked with the shorter cross-linking agent.

CONCLUSION

The effect of cross-linking on temperature-induced relaxation of plastic deformation of PMMA was studied. Increasing concentration of cross-linking agent in polymerization system to 5.2 mol.% is responsible for increase in low-temperature component of relaxation. For polymer samples with 11.2 mol.% of cross-linking agent, complete relaxation is observed at temperatures well below glass transition temperature via only low-temperature component. Increasing contribution from low-temperature component to total relaxation of plastic deformation, and, finally, degeneration of high-temperature component is likely to be controlled by restriction of segmental mobility within certain structural sublevels.

Increasing the length of cross-links results in decrease in low-temperature component of relaxation. This behavior may be explained by the fact, that the higher restriction of segmental mobility is observed for the samples cross-linked with the shorter cross-linking agent.

This experimental evidence suggests the possibilities to control low-temperature reversibility of strain in polymer glasses at temperatures well below glass transition temperature via their cross-linking.

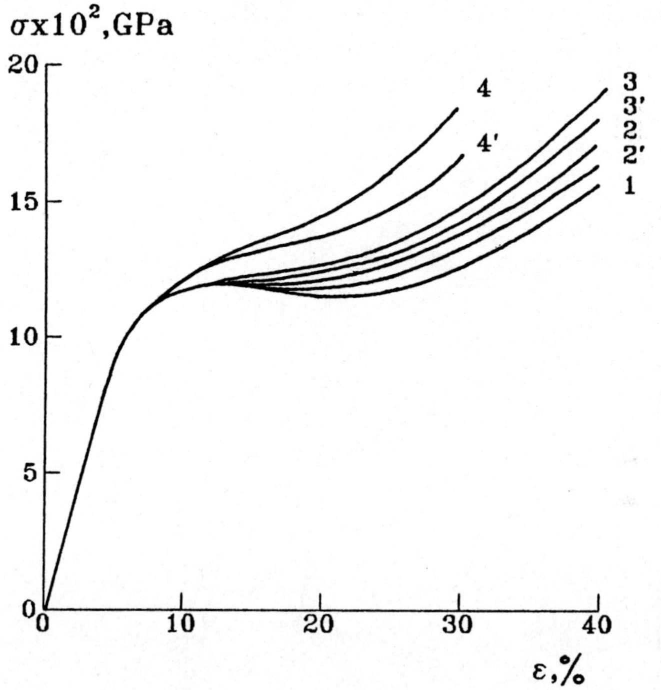

Fig. 1. Stress-strain diagrams of PMMA (1) and PMMA, cross-linked with EGDM (2 - 4) and TGDM (2' - 4'). Concentration of cross-linking agents: 2.5 (2, 2'), 5.2 (3, 3') and 11.2 mol.% (4, 4'). Uniaxial compression at 293 K.

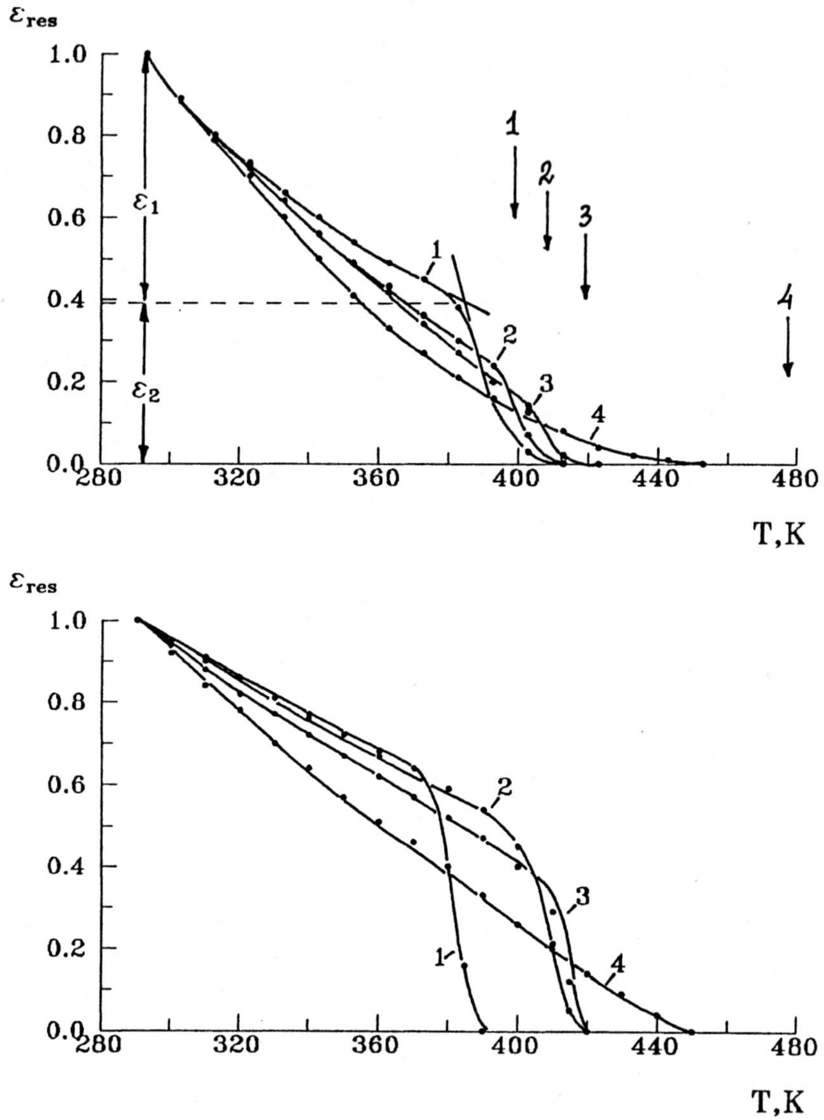

Fig. 2. Temperature dependences of residual deformation ε_{res} of PMMA (1) and PMMA, cross-linked with 2.5 (2), 5.2 (3) and 11.2 mol.% EGDM (4). Strain: 20 (A) and 30% (B). The arrows show T_g of the corresponding polymer samples.

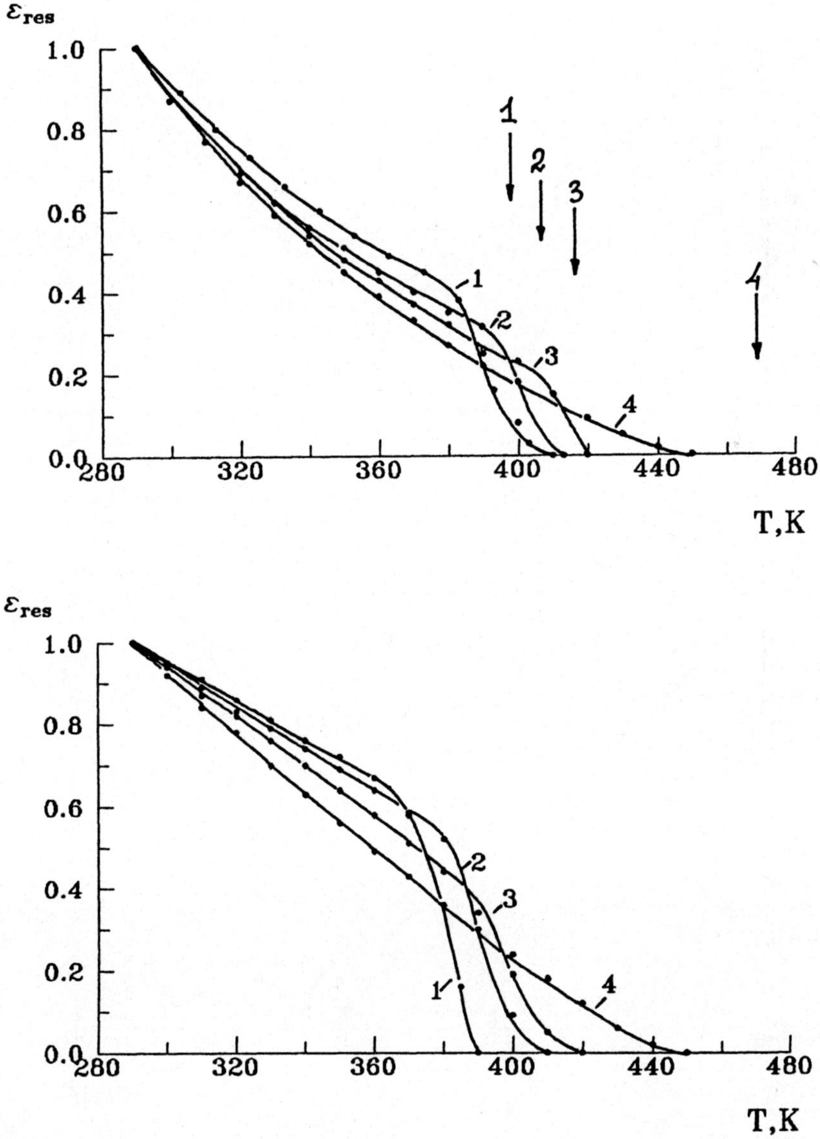

Fig. 3. Temperature dependences of residual deformation ε_{res} of PMMA (1) and PMMA, cross-linked with 2.5 (2), 5.2 (3) and 11.2 mol.% TGDM (4). Strain: 20 (A) and 30% (B). The arrows show T_g of the corresponding polymer samples.

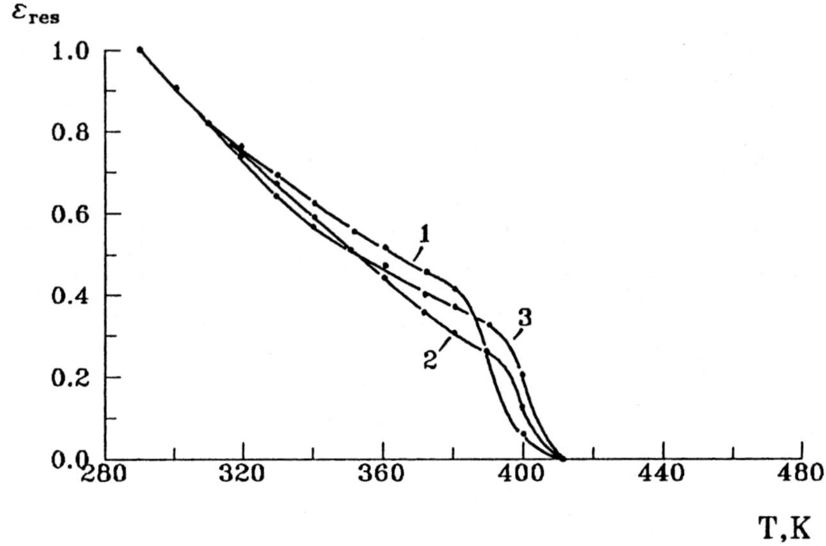

FIG. 4A

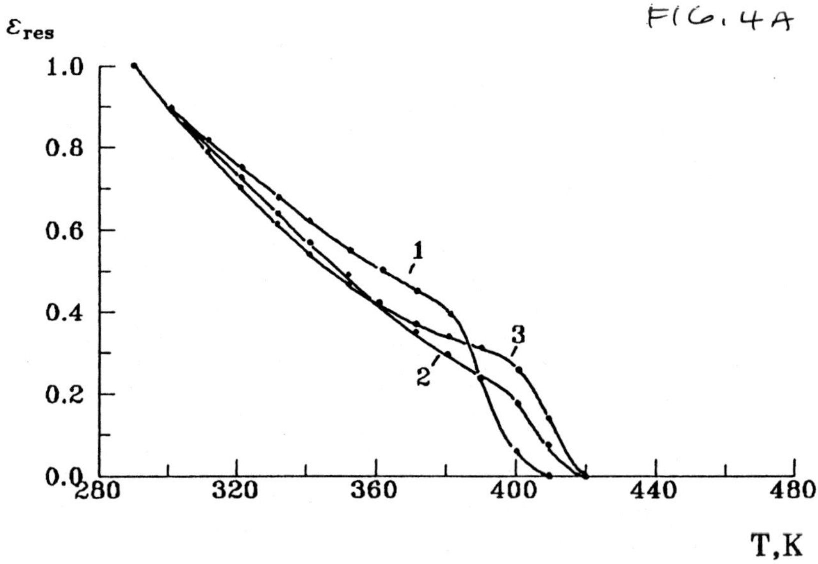

Fig. 4. Temperature dependences of residual deformation ε_{res} of PMMA (1) and PMMA, cross-linked with EGDM (2) and TGDM (3). Concentration of cross-linking agents: 2.5 (A) and 5.2 mol.% (B). Strain: 20%.

REFERENCES

1. S.A. Arzhakov, Doctorate (Chem.) Dissertation, Moscow: Karpov Institute of Physical Chemistry, 1975 (in Russian).

2. S.A. Arzhakov, N.F. Bakeev and V.A. Kabanov, Vysokomol. Soedin., A, 15, 1154 (1973) (in Russian).

3. E. Oleynik, Progr. Colloid & Polym. Sci., 80, 140 (1989).

4. E.F. Oleinik, O.B. Salamatina, S.N. Rudnev and S.V. Shenogin, Polym. Sci., 35, 1532 (1993).

5. M.S. Arzhakov and S.A. Arzhakov, Intern. J. Polymeric Mater., 29, 249 (1995).

6. M.S. Arzhakov and S.A. Arzhakov, in: Fizikokhimiya polimerov (Physics and Chemistry of Polymers), Tver', 2, 64 (1996) (in Russian).

7. M.S. Arzhakov, S.A. Arzhakov, M.B. Lachinov, E.V. Nechaeva and D.V. Trachenko, Intern. J. Polymeric Mater., (1996) (in press).

8. V.P. Shantarovich, M.S. Arzhakov and I.B. Kevdina, Polym. Sci., B, 37, 171 (1995).

9. M.S. Arzhakov, S.A. Arzhakov, I.B. Kevdina and V.P. Shantarovich, in: Fizikokhimiya polimerov (Physics and Chemistry of Polymers), Tver', 2, 59 (1996) (in Russian).

10. V.I. Irzhak, B.A. Rozenberg and N.S. Enikolopyan, Setchatye polimery (Polymer Networks), Moscow: Nauka, 1979 (in Russian).

Physical and Mechanical Behavior of Polymer Glasses in Terms of Temperature-Induced Relaxation of Plastic Deformation

M.S. Arzhakov and S.A. Arzhakov

Polymer Department, Faculty of Chemistry, Lomonosov Moscow State University, Vorob'evy Gory, Moscow, 119899 Russia

Temperature-induced relaxation of residual plastic deformation of polymer glasses was discussed in terms of structural inhomogeneity of glassy polymers. Mechanical properties of glassy polymers were shown to be controlled by the ratio between low-temperature and high-temperature components of temperature-induced relaxation of residual deformation. This ratio was treated as a new physical parameter describing physical and mechanical behavior of polymer glasses. The applied aspects of studying temperature-induced relaxation of residual deformation concerning the prepartion of modern polymer materials were discussed.

Key words: polymer glasses, plastic deformation, relaxation, mechanical properties

INTRODUCTION

Polymer glasses are widely used as modern materials (for example, engineering plastics) because of their unique properties, in particular, lowered brittleness and well-pronounced ductility as compared with low-molecular-mass glassy materials. For further development of advanced polymeric materials based on glassy polymers one should understand the origin of the specific properties of polymer glasses and, in general, the origin of their physical and mechanical behavior.

To solve this problem, studying temperature-induced relaxation of residual plastic deformation in polymer glasses was shown [1 - 6] to provide an important information. Temperature-induced relaxation of plastic deformation in polymer glasses is known [7 - 10] to involve low-temperature component at temperatures well below glass transition temperature T_g and high-temperature component at temperatures close to T_g. Obviously, such complicated character of relaxation of plastic deformation is a manifestation of physical processes, which are responsible for and accompany plastic deformation. In this case, development of plastic deformation and physical and mechanical behavior of glassy polymers are likely to be controlled by the coexistence of the above components of relaxation.

Earlier [4, 5, 11], mechanical characteristics and the ratio between low-temperature and high-temperature components of temperature-induced relaxation of plastic deformation of polymer glasses were shown to be well correlated. The origin of this correlation was discussed in terms of structural inhomogeneity of polymer glasses.

In the present paper, we summarized our previous speculations concerning the description of physical and mechanical behavior of polymer glasses from the standpoint of temperature-induced relaxation of plastic deformation. These speculations are based on the studying temperature-induced relaxation of plastic deformation in virgin PMMA [4, 5, 10], PMMA plasticized with dibutyl phtalate (DBPh) [4, 5], PMMA containing diethyl siloxane oligomer (DES) [12, 13], PMMA cross-linked with ethylene glycol dimethacrylate (EGDM) and triethylene glycol dimethacrylate (TGDM) [14], and copolymers of methyl methacrylate (MMA) with butyl methacrylate (BMA), octyl methacrylate (OMA), and lauryl methacrylate (LMA) [11].

1. Some Comments Concerning the Experimental Methods to Study the Temperature-Induced Relaxation of Residual Plastic Deformation of Polymer Glasses.

To study temperature-induced relaxation of residual deformation of polymer glasses we used the following procedures:

A. For initial PMMA, PMMA plasticized with DBPh, and PMMA containing DES, cylindrical test samples with a height h_0 were uniaxially compressed at different temperatures to a given strain. After that, they were cooled with liquid nitrogen in a stressed state. Then, the free-standing samples were heated to 213 K. The height of the samples at this temperature was denoted as h_{213}. The relaxation of residual deformation was estimated by measuring the height of the samples on heating from 213 K at a heating rate of 1 K/min. The height of the samples at a given temperature was denoted as h_T. The relative residual deformation at a given temperature was estimated as $\varepsilon_{res}=(h_0-h_T)/(h_0-h_{213})$. For PMMA sample, typical temperature dependence of ε_{res} is shown in Fig. 1. Low-temperature component ε_1 and high-temperature component ε_2 of residual deformation was estimated as follows from Fig.1. Note, that, in this case, a certain fraction of residual deformation is lost on heating free-standing polymer sample to initial test temperature (213 K).

B. For copolymers of MMA with BMA, OMA, and LMA, cylindrical test samples with a height h_0 were uniaxially compressed at 293 K to a given strain. The height of stressed samples was denoted as h_{def}. After that, they were unloaded at this temperature. The free-standing samples were heated to a given temperature and allowed to relax at this temperature until the equilibrium height h_T was attained. Relative residual deformation at a given temperature was estimated as $\varepsilon_{res}=(h_0-h_T)/h_0-h_{def})$. Note, that, in this case, residual deformation estimated involves Hookean component of relaxation.

Reference tests demonstrated that the results obtained by methods A and B are in a fair agreement and may be compared with each other.

C. For cross-linked PMMA samples, cylindrical test samples with a height h_0 were uniaxially compressed at 293 K to a given strain. After that, they were unloaded at this temperature and allowed to relax at this temperature until the equilibrium height h_{293} was attained. Then, the free-standing samples were heated from 293 K at a heating rate of 1 K/min. The height of the samples at a given temperature was denoted as h_T. Relative residual deformation at a given temperature was estimated as $\varepsilon_{res}=(h_o-h_T)/h_o-h_{293})$. In this case, the fraction of residual deformation, which is able to relax at temperatures below 293 K, is lost. So, the results obtained by this method can not be quantitatively compared with those obtained by methods A and B.

2. Structural Interpretation of Temperature-Induced Relaxation of Plastic Deformation of Polymer Glasses

A typical temperature dependence of relative residual deformation ε_{res} in PMMA samples (Fig. 1) shows that residual deformation ε_{res} involves low-temperature (ε_1) and high-temperature (ε_2) components and may be represented as follows

$$\varepsilon_{res} = \varepsilon_1 + \varepsilon_2 \qquad (1)$$

Note, that temperature-induced relaxation of residual deformation via high-temperature component ε_2 is observed in rather narrow temperature region close to glass transition. Relaxation of plastic deformation via low-temperature component ε_1 takes place in a wide temperature region well below glass transition temperature, and each temperature is associated with a certain portion of relaxation of residual deformation.

High-temperature component of relaxation is associated with the transition of excited chain conformations to their initial equilibrium state via segmental mobility in polymer sample. The origin of low-temperature component of relaxation is not so clear. At the present time, the appearance of this mode of relaxation is treated in terms of two approaches: recovery of small-scale plastic shear transformations [1, 2] (i) and structural inhomogeneity of glassy polymers [4, 5, 10] (ii). Within the framework of the latter approach, glassy polymer is characterized by a set of structural sublevels with different packing densities and ordering.

Each of these sublevels is associated with its own relaxation time and activation energy of segmental mobility. As a result, at each temperature below glass transition temperature, local segmental mobility in certain structural sublevels is allowed. With increasing temperature, structural sublevels with higher packing densities acquire segmental mobility, and finally at T_g, all polymer sample is involved in segmental mobility. Hence, a wide spectrum of segmental mobility in glassy polymer is provided by the existence of a set of structural sublevels with different packing densities and local ordering.

In the case of the deformed polymer samples, as temperature increases, the appearance of segmental mobility in structural levels with higher packing density is accompanied by a gradual recovery of local plastic deformation within the above structural sublevels. In other words, low-temperature relaxation of residual deformation is controlled by low-temperature fraction of the spectrum of segmental mobility because of temperature-induced segmental mobility within structural sublevels with lower packing densities at temperatures well below T_g. A complete relaxation of plastic deformation at T_g via high-temperature component is controlled by high-temperature fraction of the spectrum of segmental mobility. Hence, molecular mechanism of low-temperature relaxation is similar to that of high-temperature mode and involves transition of excited chain conformations to initial equilibrium state via segmental motion within certain structural sublevels.

In [4, 5], the ratio between the components associated with low-temperature and high-temperature relaxation was studied as a function of strain. For the compressed samples with rather low strains (Fig. 2, curves 1, 2), complete relaxation takes place at temperatures well below T_g. On the contrary, a complete relaxation of the compressed samples with high strains is observed on heating to T_g (Fig. 2, curves 3 - 5). Note, that in this case, low-temperature component of relaxation is also involved.

Figure 3 presents low-temperature (ε_1) and high-temperature (ε_2) components of temperature-induced relaxation of plastic deformation as a function of strain and the corresponding stress-strain curve of uniaxial compression of PMMA samples. As follows from Fig. 3, in the case of the samples with strains below yield strain ε_y, temperature-induced relaxation of plastic deformation proceeds via low-temperature component ε_1. At the strain corresponding to ε_y, component ε_1 is maximal, and further increase in compression strain is associated with the appearance and growth in the high-temperature component ε_2 at constant ε_1.

The results obtained allowed us to advance the following structural pattern of the development of plastic deformation in glassy polymers. At strains below yield strain, deformation is controlled by the stress-induced segmental mobility within structural sublevels with lower packing densities. At temperatures well below T_g, temperature-induced relaxation of this portion of plastic deformation is controlled by low-temperature mode of the spectrum of segmental mobility. At strains above yield strain, plastic deformation is controlled by the stress-induced segmental mobility within densely packed structural sublevels. This portion of plastic deformation is able to relax at temperatures close to T_g, when all polymer sample is involved in segmental mobility.

Let us discuss the validity of the mechanism proposed, namely, the validity of the following assumptions: structural inhomogeneity of polymer glasses (i), and realization of low-temperature segmental mobility at temperatures well below T_g within certain structural sublevels (ii).

2.1. On Structural Inhomogeneity of Polymer Glasses.

Note, that during recent decades, structural inhomogeneity in glassy polymers is widely discussed in literature. Review of existing experimental evidences is beyond the scope of the present paper. Let us mention only some of them.

As demonstrated by electron microscopy [15 - 22], electron diffraction [21, 23], positron annihilation lifetime spectroscopy [12, 24], light scattering measurements [25], small-angle X-ray scattering [19], and wide-angle X-ray scattering [23], polymer glasses are characterized by structural inhomogeneity related to the existence of heterogeneous ordered regions with higher packing density (domains, clusters) with dimensions up to several nanometers. These domains are joined with each other by numerous tie chains, which constitute the regions with lower packing densities and degrees of ordering.

Experimental evidence obtained provided a fundamental basis for the development of various structural models of glassy polymers. The difference between the models advanced is related to the type of principal morphological structural unit: fibril [9, 10], domain [26, 27], globule [28 - 30], and cluster [31, 32]. However, at the present time a certain lack of convincing experimental evidence does not allow one to decide in favor of this or that particular model. Hence, in this work, our speculations will be based only on unequivocal conclusion concerning structural inhomogeneity of glassy polymer, which is associated with the existence of, at least, two local structural sublevels with different packing densities and degrees of ordering of macromolecules or their fragments. Note, that these speculations allow a plausible description of specific effect of plasticizing agents on molecular mobility in polymers [33, 34], dual-mode sorption of gases by glassy polymers [35 – 37], and development of the irreversibly bound form of low-molecular-mass compounds in polymers [38 - 40].

In our opinion, the speculations concerning structural inhomogeneity in polymer glasses should have logical development. Structural sublevels with higher and lower packing densities as recognized by modern structural studies are limiting cases in structural organization of glassy polymer. We believe, that there is a continious spectrum of intermediate states (structural sublevels) with different packing densities and degrees of ordering. As was mentioned above, this structural spectrum is likely to be responsible for the spectrum of segmental mobility in glassy polymers.

2.2. On Low-temperature Segmental Mobility in Polymer Glasses.

A possible realization of segmental mobility in local structural sublevels at temperatures well below glass transition temperature is widely discussed in literature. Let us mention some works [34, 41 - 43], in which specific dual transitions observed in temperature region of glass transition were treated in terms of existence of structural regions with different packing densities. In this case, low-temperature transition is attributed to the appearance of segmental mobility in loosely packed regions, whereas high-temperature transition is related to segmental mobility in densely packed domains.

Speculations concerning the mechanism of the β-relaxation in glassy polymers [44] based on DSC measurements of activation parameters of the β-transition [45 - 47] allowed one to conclude that β-transition involves quasi-independent and localized motion of statistic segments within structural regions with lower packing densities.

As was shown in [12, 24] using positron annihilation lifetime spectroscopy, the dimensions of free volume microregions in loosely packed structural regions in PMMA at 293 K are equal to 0.3 - 0.4 nm^3. By order of magnitude, this value coincides with activation volume of relaxation segmental transition (0.8 - 1.0 nm^3) and with the volume of statistic segment of PMMA (0.9 nm^3) [44]. This fact may serve as a supporting evidence concerning realization of segmental mobility within structural sublevels with lower packing densities at temperatures well below T_g.

Additional factors assisting segmental mobility in local structural regions in glassy polymer are the following. First, asymmetry of the field of intermolecular interaction provided by free volume microregions is associated with a decrease in activation energy of segmental mobility [48]. Secondly, a certain decrease in activation energy of segmental mobility may be also related to the existence of inner stresses stored in polymer during deformation [9, 10]. From this standpoint, low-temperature segmental mobility is controlled by mechanical activation under inner stresses stored in deformed polymer sample.

Hence, literature data evidence the validity of our speculations concerning the structural pattern of plastic deformation and its temperature-induced relaxation in polymer glasses.

To provide experimental support of the mechanism proposed, the structural changes during plastic deformation were studied using positron annihilation lifetime spectroscopy [12, 24]. Application of this technique allows one to estimate the effective radius R_+ and concentration of free volume microregions N_+ in structural sublevels with higher packing densities as well as the corresponding values of R_{Ps} and N_{Ps} in structural sublevels with lower packing densities. As was shown, for initial PMMA samples at 293 K, $R_+ \approx 0.34$ nm, $N_+ \approx 0.7 \times 10^{16}$ cm^{-3}, $R_{Ps} \approx$ 0.42 nm, and $N_{Ps} \approx 1.46 \times 10^{19}$ cm^{-3}. For plastically deformed PMMA samples (residual deformation stored in unloaded samples was 16.5%), the corresponding values were the following: $R_+ \approx 0.34$ nm, $N_+ \approx 1.0 \times 10^{16}$ cm^{-3}, $R_{Ps} \approx 0.45$ nm, and $N_{Ps} \approx 0.66 \times 10^{19}$ cm^{-3}. As follows from these data, deformation of PMMA is accompanied by slight change in N_+ from 0.7×10^{16} to 1.0×10^{16} cm^{-3}. Note, that this change in N_+ lies within anticipated experimental error (20 %). At the same time, a two-fold decrease in N_{Ps} is observed: from 1.46×10^{19} to 0.66×10^{19} cm^{-3}. Dimensions of free volume microregions (R_+ and R_{Ps}) remains unchanged. This experimental evidence allows one to conclude, that, for PMMA samples with strains close to yield strain, deformation is likely to be controlled by densification of structural regions with lower packing densities whereas densely packed structural sublevels remain intact.

Obviously, any changes in initial polymer structure and concomitant changes in packing density within the above structural sublevels should have a certain effect on both mechanical and relaxation properties of glassy polymers. Structural changes in polymer may be accomplished by introduction of incompatible low-molecular-mass compounds, which are unable to produce a uniform distribution in polymer volume. This approach was formalized in terms of structural plasticization of polymers [49, 50]. Structural plasticization may be rationalized as follows: a low-molecular-mass compound with a low thermodynamic affinity to polymer is localized within local structural regions at boundaries between supramolecular or suprasegmental structures of glassy polymer. In this case, within these regions, changes in mobility of macromolecules or their fragments is observed, and this has a certain effect on mechanical behavior of polymer material. In [13], for PMMA, diethyl siloxane oligomer (DES) was used as such low-molecular-mass modifying agent.

Figure 4 shows the concentration dependences of yield stress σ_y for uniaxial compression of PMMA samples containing DES (curve 1) as well as contribution from low-temperature component to total temperature-induced relaxation of residual deformation $\varepsilon_1/(\varepsilon_1+\varepsilon_2)$ (curve 2). As follows from Fig. 4, at 0.1 wt.% of DES in PMMA, an increase in both σ_y and $\varepsilon_1/(\varepsilon_1+\varepsilon_2)$ is observed. With increasing concentration of DES in polymer, these parameters decrease. The effect of DES in PMMA on σ_y and $\varepsilon_1/(\varepsilon_1+\varepsilon_2)$ is likely to be related to changes in segmental mobility in certain structural sublevels because of changes in packing density and fractional content of free volume. To verify this assumption, the positron annihilation lifetime spectroscopy was used to follow changes on elementary free volume induced by introduction of DES in PMMA samples.

For all samples studied at 293 K, $R_+ \approx 0.34$ nm and $R_{Ps} \approx 0.42$ nm. For initial PMMA and PMMA containing 0.1 wt.% of DES, $N_+ \approx 0.9\text{x}10^{16}$ cm^{-3} and $N_{Ps} \approx 1.0\text{x}10^{19}$ cm^{-3}. As content of DES in PMMA increases, increasing both N_+ and N_{Ps} is observed. For PMMA samples containing 0.5 and 1.3 wt.% of DES, $N_+ \approx 1.15\text{x}10^{16}$ cm^{-3}

and $N_{Ps} \approx 2.2x10^{19}$ cm^{-3}.

Hence, introduction of DES into PMMA is not accompanied with the changes in the dimensions of free volume microregions both in structural regions with increased and decreased packing density in polymer. The content of free volume microregions in more densely packed structural regions N_+ changes with increasing DES content in PMMA but still lies within expected error (20%). Concentration of free volume microregions in loosely packed structural regions N_{Ps} increases by two times, and this evidence suggests that introduction of DES into polymer is associated with loosening of these structural regions. Constancy of annihilation characteristics for PMMA samples containing 0.1 wt % of DES as compared with virgin PMMA may be rationalized by the fact that low-molecular-mass additive occupies rather large free volume microregions. These regions are located far away from each other, and distances between them are much higher than diffusion paths of positron and positronium. As a result, they appear to be beyond the detection limit of this method.

This structural interpretation of experimental evidence obtained by the method of positron annihilation correlates well with earlier structural studies in polymer glasses (for example, [15 - 17]). Structure of glassy polymer was shown to be characterized by a set of densely packed regions (domains) separated by structural regions with lower packing density. (Let us note that the term "domains" is used only to specify structural elements with increased packing density and carries no information concerning the structure of these regions.) Domains are able to aggregate with each other and produce more complicated supradomain structures. The boundary regions between supradomain structures are more loosely packed as compared with interdomain regions. Hence, experimental evidence obtained allowed identification of three structural sublevels in glassy polymers listed in order of increase in free volume: domains, interdomain regions, and boundary regions between supradomain structures. Three-component treatment of lifetime distributions of annihilation radiation allows one to obtain information concerning free volume fraction for domains and interdomain regions.

According to this approach, the effect of introduction of DES on mechanical properties of PMMA containing DES may be presented as follows. At low content of DES (0.1 wt %), DES molecules occupy rather large free volume microregions located at boundaries between supradomain structures. As a result, segmental mobility in the vicinity of these microregions is hindered, and mechanical properties of polymer material increase (Fig. 4). With increasing the content of DES in PMMA to 0.5 and 1.3 wt %, low-molecular-mass additive is able to enter interdomain structural regions. This penetration is accompanied by loosening of these regions and additional free volume microregions appear. As a result, within these structural regions, segmental mobility of macromolecules is enhanced, and σ_y and $\varepsilon_1/(\varepsilon_1+\varepsilon_2)$ decrease (Fig. 4). During introduction of DES into PMMA, densely packed domains remain intact and do not contribute to changes in mechanical properties of the system.

The effect of DES on segmental mobility of PMMA is supported by DSC data (Fig. 5). At 0.1 % of DES in PMMA, endothermic jump in glass transition region increases as compared with initial PMMA (transition from curve 1 to curve 2). This evidence suggests a suppressed segmental mobility of polymer chains for polymer samples containing low-molecular-mass additive. As concentration of DES in PMMA increases (curve 3), endothermic jump decreases, and this behavior is related to an enhanced segmental mobility of polymer chains.

Hence, structural interpretation of temperature-induced relaxation of plastic deformation allowed us to advance structural pattern of plastic deformation of polymer glasses based on gradual involvement of segmental mobility within structural sublevels with different packing densities in deformation and relaxation. The validity of this structural pattern is supported by literature data as well as by direct structural studies using positron annihilation lifetime spectroscopy.

3. A New Approach to Description of Physical and Mechanical Behavior of Polymer Glasses.

As was mentioned above, low-temperature relaxation of residual deformation is controlled by low-temperature fraction of the spectrum of segmental mobility because of temperature-induced segmental mobility within structural sublevels with lower packing densities at temperatures well below T_g. Evidently, a certain part of low-temperature fraction of the spectrum of segmental mobility is able to relax during polymer loading. Let us define this component as ε_1^*. From this standpoint, low-temperature component ε_1 of relaxation is likely to be associated with stored plastic deformation, which remained unrelaxed during loading.

High-temperature fraction of the spectrum of segmental mobility is responsible for high-temperature component ε_2 of relaxation. This component is accumulated in polymer during loading, and is able to relax only at temperatures close to T_g via temperature-induced segmental mobility within structural sublevels with higher packing densities.

From this standpoint, plastic deformation of glassy polymer may be presented as

$$\varepsilon = \varepsilon_1^* + \varepsilon_{res} = (\varepsilon_1^* + \varepsilon_1) + \varepsilon_2 \qquad (2)$$

According to equation (2), plastic deformation of glassy polymer may be described by the contributions from components ε_1 and ε_2 to temperature-induced relaxation of residual deformation. At $\varepsilon \le \varepsilon_y$, deformation is primarily controlled by $(\varepsilon_1^* + \varepsilon_1)$, whereas at $\varepsilon \ge \varepsilon_y$ component ε_2 comes into play.

With respect to polymer strength properties, consideration of component $(\varepsilon_1^* + \varepsilon_1)$ is of most interest. This parameter is responsible for the development of polymer deformation at initial portion of $\sigma - \varepsilon$ curve, i.e., at $\varepsilon \le \varepsilon_y$, which is associated with such mechanical properties as elastic modulus E_o and yield stress σ_y. This allows one to suggest the correlation between relaxation $(\varepsilon_1^* + \varepsilon_1)$ and mechanical characteristics E_o and σ_y.

To verify this correlation, PMMA samples were compressed to different strains at different temperatures. Then, temperature-induced relaxation of residual deformation ε_{res} was studied (Fig. 6). For the lack of space, recovery curves are presented only for the samples with 14 and 20% strains. As deformation temperature T_{def} increases, the contribution from low-temperature component ε_1 to relaxation of residual deformation is seen to decrease. At test temperatures close to glass transition temperature, the relaxation of ε_{res} is controlled only by high-temperature component ε_2 (Fig. 6, curves 5). On the other hand, as T_{def} increases, elastic modulus and yield point go down.

This experimental evidence suggests an intimate correlation of elastic modulus and yield stress with the contribution from low-temperature component to total relaxation of plastic deformation $\varepsilon_1/(\varepsilon_1+\varepsilon_2)$ (Fig. 7). At any strains, similar tendency is observed: as contribution from component ε_1 to relaxation of ε_{res} decreases, both elastic modulus and yield stress tend to decrease.

With increasing T_{def}, the parallel decrease in both $\varepsilon_1/(\varepsilon_1+\varepsilon_2)$ and E_0 and σ_y is likely to be controlled by the fact, that, in this case, local segmental mobility in certain structural sublevels is enhanced. As a result, the higher fraction of low-temperature mode of the whole spectrum of segmental mobility is able to relax during deformation. In terms of equation (2), this implies an increase in component ε_1^* via decreasing component ε_1. In turn, as contribution from component ε_1 to relaxation of ε_{res} decreases, component ε_2 increases.

Obviously, the enhancement of segmental mobility in polymer

sample is controlled not only by increasing T_{def}, but also by decreasing T_g. Depression in T_g may be attained by either plasticization of PMMA or by copolymerization of MMA with methacrylic monomers. To verify this idea, for PMMA samples plasticized with DBPh as well as for copolymers of MMA with BMA, OMA, and LMA, mechanical behavior and temperature-induced relaxation of residual deformation were studied [4, 5, 11]. As was shown, with increasing concentration of DBPh in PMMA and increasing concentration of BMA, OMA, and LMA in copolymers with MMA, decrease in both σ_y and $\varepsilon_1/(\varepsilon_1+\varepsilon_2)$ is observed (Table 1).

For copolymers of MMA as well as for PMMA deformed at different T_{def} and plasticized PMMA, Fig. 8 shows the dependence of σ_y on $\varepsilon_1/(\varepsilon_1+\varepsilon_2)$. As follows from Fig. 8, all experimental data fit one curve. Hence, there is a good correlation between σ_y and $\varepsilon_1/(\varepsilon_1+\varepsilon_2)$ and this correlation is independent of any variations in these parameters: either by varying T_{def} of PMMA, or plasticization of PMMA, or copolymerization of MMA with BMA, OMA, and LMA. This fact is likely to be associated with the difference between T_g and T_{def}, which is defined as $\Delta T_{def} = T_g - T_{def}$.

Experimental evidence presented in Table 1 allows one to estimate both σ_y and $\varepsilon_1/(\varepsilon_1+\varepsilon_2)$ at a certain ΔT_{def} (Table 2). At a given ΔT_{def}, σ_y and $\varepsilon_1/(\varepsilon_1+\varepsilon_2)$ remain unchanged and are independent of any variations in ΔT_{def}: by increasing T_{def} at a fixed T_g or by decreasing T_g at a fixed T_{def}. Note, that such behavior is likely to be associated with the dynamics of segmental mobility in polymer glasses.

The dynamics of segmental mobility seems to be controlled not only by ΔT_{def} but also by strain rate [51, 52]. One could expect, that with increasing strain rate, the fraction of low-temperature mode of segmental mobility, which is able to relax during deformation, decreases. As a result, both σ_y and $\varepsilon_1/(\varepsilon_1+\varepsilon_2)$ should increase. To verify this idea, copolymers of MMA with BMA, OMA, and LMA were tested under the compression with different strain rates.

For copolymer samples, with increasing strain rate from 0.1 to 100 %/min, increase in both σ_y and $\varepsilon_1/(\varepsilon_1+\varepsilon_2)$ is observed. Figure 9 shows the dependence of σ_y on $\varepsilon_1/(\varepsilon_1+\varepsilon_2)$, when these parameters were varied by changing strain rate. In this case, a well-pronounced correlation between σ_y and $\varepsilon_1/(\varepsilon_1+\varepsilon_2)$ is also observed.

The above experimental evidence allows one to suggest that restriction of segmental mobility in polymer should result in the growth in contribution from low-temperature component ε_1 to relaxation. Note, that such restriction of segmental mobility in polymer sample may be achieved by cross-linking of polymers [53] To verify this idea, for PMMA samples cross-linked with bifunctional monomers ethylene glycol dimethacrylate (EGDM) and triethylene glycol dimethacrylate (TGDM), temperature-induced relaxation of residual deformation was studied.

For PMMA samples, cross-linked with EGDM, temperature dependences of residual deformation are shown in Fig. 10. Note, that, with increasing concentration of cross-inking agent in PMMA samples to 2.5 and 5.2 mol% (curves 2 and 3), contribution from low-temperature component ε_1 to total relaxation of residual deformation increases as compared with initial PMMA (curve 1). With increasing concentration of EGDM to 11.2 mol% (curve 4), high-temperature component ε_2 is degenerated, and temperature-induced relaxation of residual deformation proceeds via low-temperature component ε_1. Complete relaxation is observed at temperatures well below T_g within glassy state.

Hence, for cross-linked PMMA samples, restriction of segmental mobility within certain structural sublevels is accompanied by increasing contribution from low-temperature component ε_1 to total temperature-induced relaxation of residual deformation. This experimental evidence may be attributed to the fact, that, in this case, low-temperature mode of the whole spectrum of segmental mobility is not able to relax during deformation. Obviously, restriction of segmental mobility should be controlled by the length of the molecule of cross-linking agent.

Figure 11 shows the temperature dependences of residual deformation ε_{res} for cross-linked PMMA samples with the similar concentration of EGDM and TGDM. At transition from EGDM to TGDM (curves 2 and 3, respectively), increasing the length of cross-links results in decrease in low-temperature component ε_1. This behavior may be attributed to the fact, that the higher restriction of segmental mobility is observed for the samples cross-linked with the shorter cross-linking agent.

Hence, the ratio between components ε_1 and ε_2 to recovery of plastic deformation provides an important information concerning specific features of polymer deformation and relaxation and is responsible for polymer mechanical behavior. This ratio may be treated as new physical parameter describing physical and mechanical behavior of polymer glasses.

Correlation between the above components of temperature-induced relaxation of residual deformation and polymer brittleness is quite evident. As was shown earlier, as T_{def} decreases, component ε_2 tends to vanish (Fig. 6). Brittle fracture of polymer material is likely to take place at temperatures, where component ε_1 is not able to store during deformation. Degeneration of component ε_1 implies that segmental mobility in polymer samples is ceased at all levels of structural inhomogeneity. Hence, temperature region from T_{br} to T_g may be treated as temperature interval, where higher limit is associated with manifestation of segmental mobility in whole polymer sample, and lower limit corresponds to frozen-in segmental mobility in whole polymer sample.

4. Applied Aspects of Studying Temperature-Induced Relaxation of Plastic Deformation of Polymer Glasses.

Well pronounced plastic deformation of polymer glasses allows the processability of these materials into final articles by rolling, stamping, pressing, etc. and a substantial improvement in their mechanical properties. One of the approaches to improve the mechanical properties of commercial glassy polymers, in particular, toughness, is provided by the method of orientation strengthening.

However, such plastically deformed or oriented materials have one main disadvantage: as operating temperature increases, they appear to lose their shape and orientation and, as a result, their mechanical properties deteriorate dramatically and approach those characteristic of undeformed and unoriented materials. As was shown above, these processes may be rationalized as temperature-induced relaxation of plastic deformation or temperature-induced shrinkage of oriented polymers via low-temperature component ε_1 and high-temperature component ε_2. In connection with this, lowering or suppression of both components of relaxation presents a challenging task for many scientists and engineers.

A satisfactory solution of this problem offers numerous advantages. If we manage to prevent the low-temperature relaxation, this implies a substantial broadening of the temperature range of operating conditions up to T_g. However, if both components of relaxation are prevented, this will allow much progress in the development of new generation of orientation-strengthened shrinkproof polymer materials based on well-known commercial glassy polymers.

Experimental evidence concerning the relaxation of residual deformation in PMMA samples compressed at different temperatures suggests that preventing low-temperature component ε_1 of relaxation is provided by increasing deformation temperature T_{def} (Fig. 6). When polymer compression is carried out at temperatures in the vicinity of T_g, the total relaxation is primarily controlled only by high-temperature component ε_2. This approach may serve as a scientific basis for traditional technology of preparation of orientation-strengthened polymer glasses. This technology involves orientation (tensile drawing) of a polymer above its glass transition temperature and its further cooling in a stressed state. Materials and articles prepared via this method are characterized by a stability of their both shape and orientation and, hence, resulting mechanical properties up to temperatures not exceeding T_g.

The fundamentals concerning the structural pattern of plastic deformation and its further temperature-induced relaxation in polymer glasses allowed us to solve the second part of the above problem: to suppress high-temperature component of relaxation of plastic deformation via a new approach to orientation strengthening of polymers [6]. Technological application of this approach makes it possible to prepare orientation-strengthened shrinkproof polymer materials based on PMMA. Mechanical characteristics of these materials appear to be comparable to those of a conventional oriented glass produced by a traditional technological method. However, the main advantage of such materials is associated with their high thermal stability, i.e., retention of shape, orientation and, finally, mechanical characteristics in a wide temperature range and even after heating the samples up to temperatures above the softening temperature of PMMA.

CONCLUSION

Studying temperature-induced relaxation of residual plastic deformation of polymer glasses was shown to provide an important information concerning development of plastic deformation and, in general, physical and mechanical behavior of glassy polymers. Interpretation of temperature-induced relaxation of residual deformation in terms of structural inhomogeneity of glassy polymers allowed us to advance structural pattern of their plastic deformation based on gradual involvement of segmental mobility within different structural sublevels in deformation and relaxation.

Mechanical characteristics of glassy polymers and the ratio between low-temperature and high-temperature components of relaxation were shown to be well correlated. This correlation is independent of any variations in the above parameters: either by varying deformation temperature and strain rate, or by physicochemical modification of polymer - plasticization, cross-linking, and copolymerization. Such behavior is likely to be associated with the dynamics of segmental mobility within certain structural sublevels in glassy polymers. These speculations allowed us to treat the ratio between low-temperature and high-temperature components of temperature-induced relaxation of residual deformation as a new physical parameter describing physical and mechanical properties of polymer glasses.

From this standpoint, the problems concerning modification of polymer materials based on polymer glasses should be treated as controlled variation in the ratio between the above components of relaxation. This approach allowed us to advance a new principle of preparation of orientation-strengthened polymer glasses, which are characterized by no shrinkage in a wide range of operating temperatures even at temperatures above glass transition temperature.

Table 1.

Physical and mechanical parameters of PMMA and various copolymers of methyl methacrylate

Sample	T_g,K	T_{def},K	σ_y,GPa	$\varepsilon_1/(\varepsilon_1+\varepsilon_2)$	ΔT_{def},K
PMMA	393	293	0.124	0.88	100
		320	0.082	0.78	73
		333	0.070	0.72	60
		353	0.060	0.61	40
		383	0.050	0.50	10
PMMA + 5 wt% of DBPh	383	293	0.102	0.85	90
PMMA + 10 wt% of DBPh	373	-	0.086	0.78	80
PMMA + 20 wt% of DBPh	343	-	0.065	0.68	50
MMA/BMA					
95/5	383	-	0.115	0.88	90
90/10	373	-	0.110	0.88	80
80/20	363	-	0.094	0.85	70
70/30	353	-	0.083	0.82	60
50/50	333	-	0.060	0.70	40
MMA/OMA					
95/5	378	-	0.096	0.83	85
90/10	363	-	0.075	0.75	70
80/20	343	-	0.055	0.63	50
MMA/LMA					
95/5	368	-	0.090	0.78	75
90/10	353	-	0.060	0.68	60
85/15	338	-	0.050	0.57	45

Table 2

Yield stress σ_y and contribution from low-temperature component to relaxation of residual deformation $\varepsilon_1/(\varepsilon_1+\varepsilon_2)$ of PMMA and various copolymers of methyl methacrylate at a given ΔT_{def}.

ΔT_{def}, K	σ_y, GPa	$\varepsilon_1/(\varepsilon_1+\varepsilon_2)$
80 - 90	0.105 ± 0.010	0.86 ± 0.02
70 - 80	0.090 ± 0.005	0.81 ± 0.04
60 - 70	0.065 ± 0.005	0.70 ± 0.02
40 - 50	0.055 ± 0.005	0.60 ± 0.03

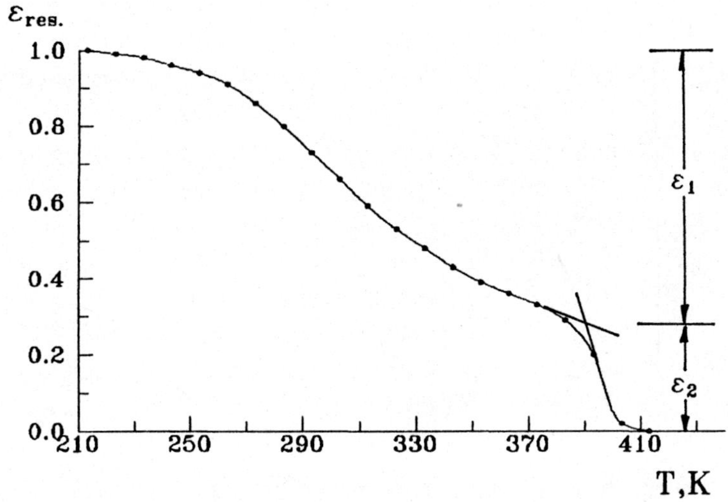

Fig. 1. Typical temperature dependence of relative residual deformation ε_{res} in PMMA sample. Strain: 20%. T_{def} = 293K.

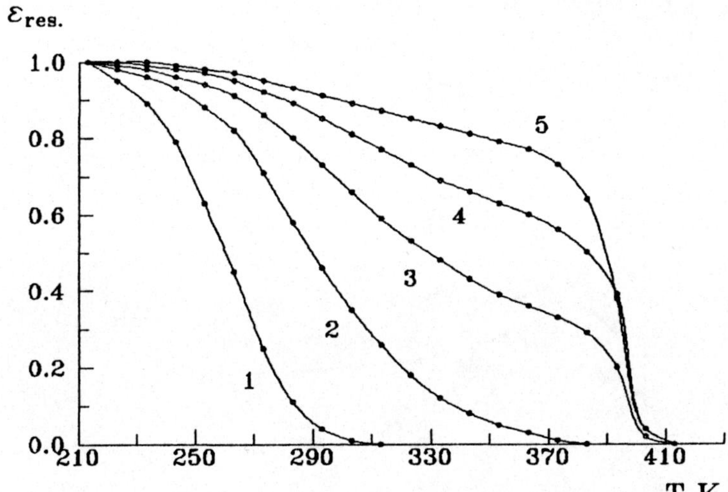

Fig. 2. Temperature dependences of relative residual deformation ε_{res} in PMMA samples with strains 10 (1), 14 (2), 20 (3), 30 (4), and 50% (5). T_{def} = 293K .

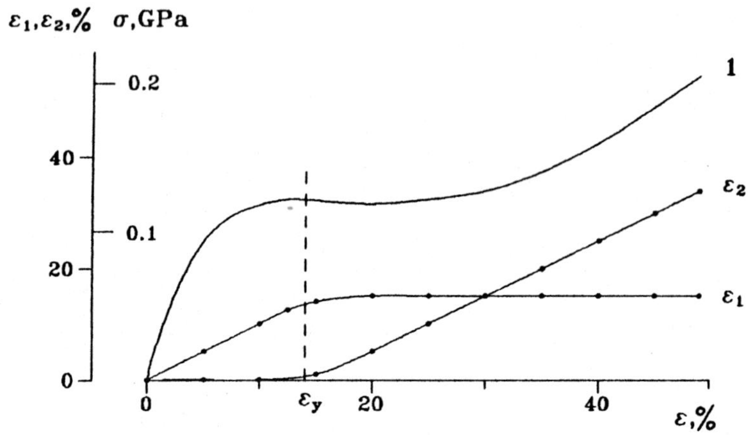

Fig. 3. Stress-strain curve corresponding to uniaxial compression of PMMA samples at 293K (1) and low-temperature ε_1 and high-temperature components ε_2 of temperature-induced relaxation of residual deformation versus strain ε.

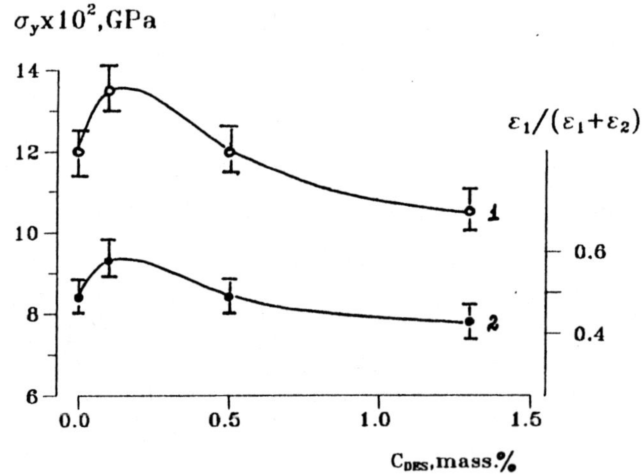

Fig. 4. Dependences of yield stress σ_y (1) and contribution from low-temperature component to relaxation of residual deformation $\varepsilon_1/(\varepsilon_1+\varepsilon_2)$ (2) on the content of DES in PMMA.

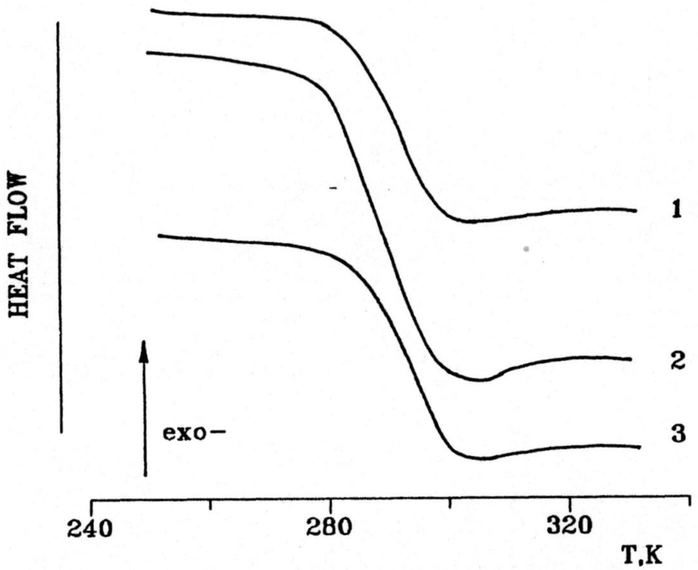

Fig. 5. DSC curves of PMMA (1) and PMMA with 0.1 (2) and 1.3 wt.% of DES (3).

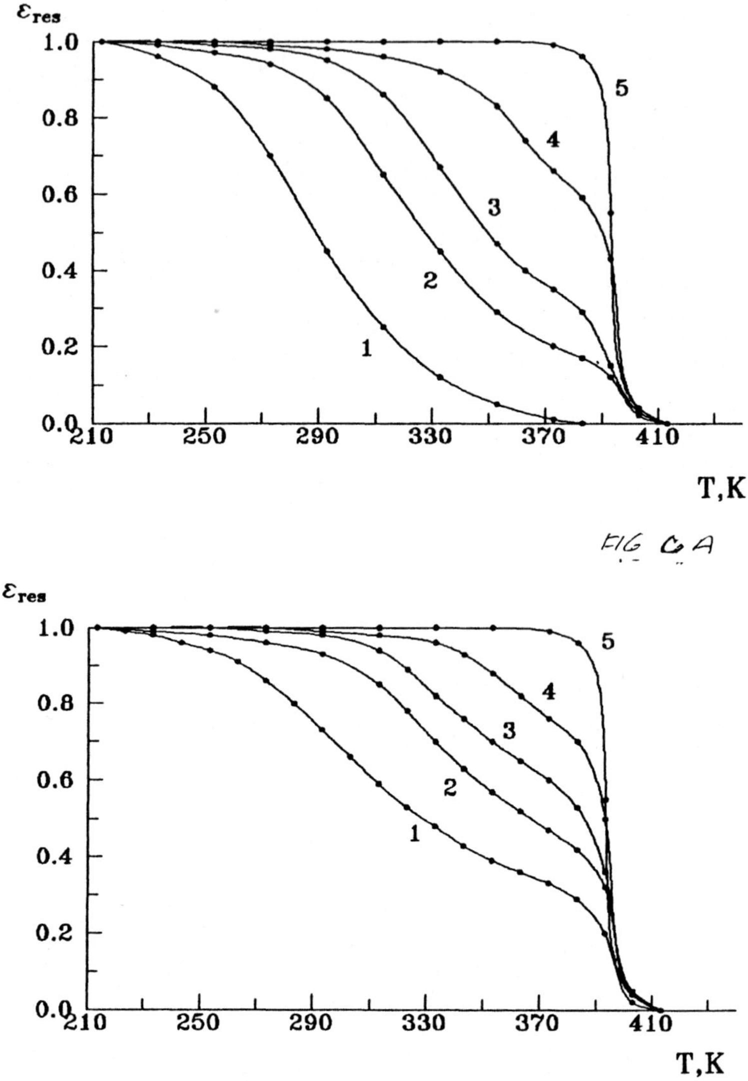

Fig. 6. Temperature dependences of relative residual deformation ε_{res} in PMMA samples at T_{def} = 293 (1), 320 (2), 333 (3), 353 (4), and 383K (5). Strain: 14 (A) and 20% (B).

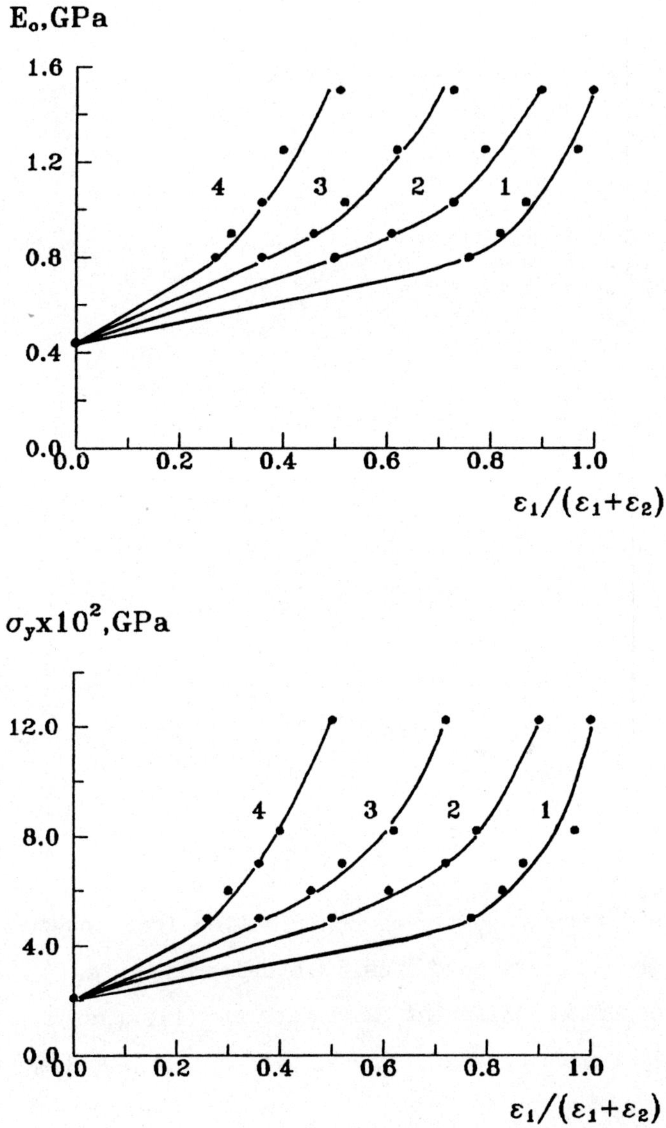

Fig. 7. The dependence of elastic modulus E_0 (A) and yield stress σ_y (B) of PMMA samples on the contribution from low-temperature component to relaxation of residual deformation $\varepsilon_1/(\varepsilon_1+\varepsilon_2)$. Strain: 10 (1), 14 (2), 20 (3), and 30% (4).

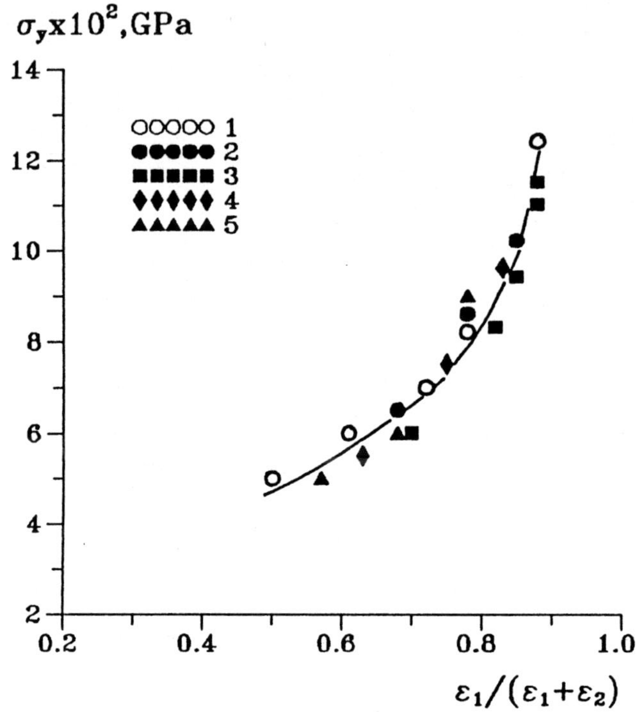

Fig. 8. Yield stress σ_y versus contribution from low-temperature component to relaxation of residual deformation $\varepsilon_1/(\varepsilon_1+\varepsilon_2)$ for PMMA compressed at different temperatures (1), plasticized PMMA (2), and copolymers of MMA with BMA (3), OMA (4), and LMA (5).

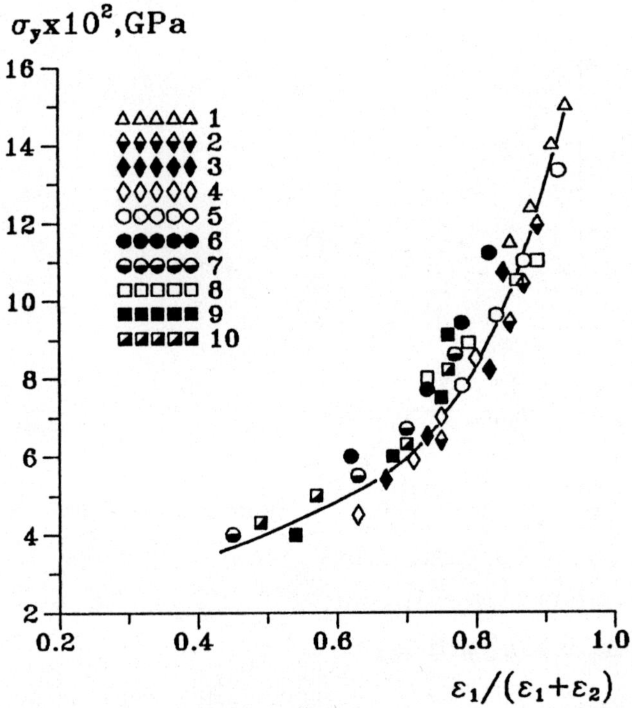

Fig. 9. Yield stress σ_y versus contribution from low-temperature component to relaxation of residual deformation $\varepsilon_1/(\varepsilon_1+\varepsilon_2)$ for PMMA (1) and various copolymers of MMA. Monomer feed composition: MMA/BMA - 80/20 (2), 70/30 (3), and 50/50 (4); MMA/OMA - 95/5 (5), 90/10 (6), and 80/20 (7); MMA/LMA - 95/5 (8), 90/10 (9), and 85/15 (10). Comments are given in text.

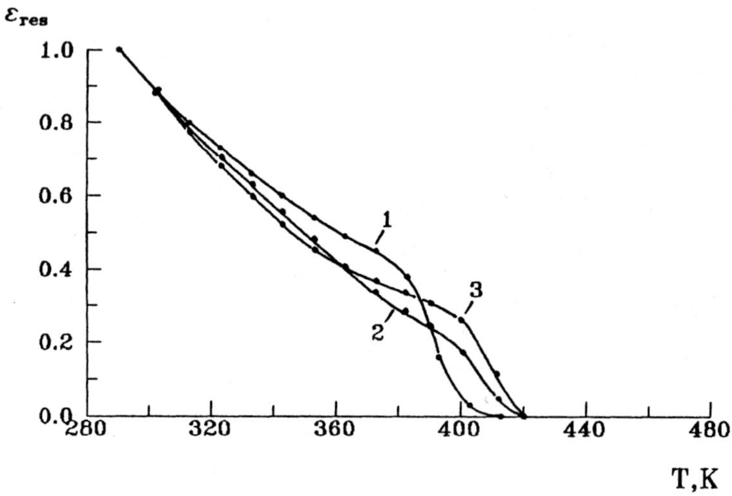

Fig. 10. Temperature dependences of relative residual deformation ε_{res} of PMMA (1) and PMMA, cross-linked with 2.5 (2), 5.2 (3), and 11.2 mol.% EGDM (4). Strain: 20 %. T_{def} = 293 K. The arrows show T_g of the corresponding polymer samples.

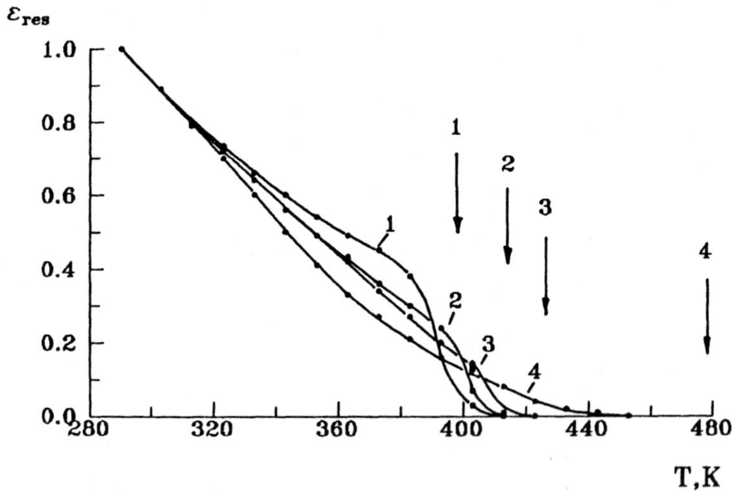

Fig. 11. Temperature dependences of relative residual deformation ε_{res} of PMMA (1) and PMMA, cross-linked with EGDM (2) and TGDM (3). Concentration of cross-linking agents: 5.2 mol.%. Strain: 20%. T_{def} = 293 K.

REFERENCES

1. E. Oleynik, Progr. Colloid & Polym. Sci., 80, 140 (1989)

2. E.F. Oleinik, O.B. Salamatina, S.N. Rudnev and S.V. Shenogin, Polymer Science, 35, 1532 (1993)

3. E.F. Oleinik, O.B. Salamatina, S.N. Rudnev and S.V. Shenogin, Polymers for Advanced Technologies, 6, 1 (1995).

4. M.S. Arzhakov and S.A. Arzhakov, Intern. J. Polymeric Mater., 29, 249 (1995).

5. M.S. Arzhakov and S.A. Arzhakov, in: Fizikokhimiya polimerov (Physics and Chemistry of Polymers), Tver', 2, 64 (1996) (in Russian).

6. M.S. Arzhakov, S.A. Arzhakov and V.A. Chernavin, New Polymeric Mater., 5, 43 (1996).

7. S.A. Arzhakov and V.A. Kabanov, Vysokomol. Soedin., B, 13, 318 (1971) (in Russian)

8. A.E. Skorobogatova, S.A. Arzhakov, N.F. Bakeev and V.A. Kabanov, Dokl. Akad. Nauk SSSR, 211, 151 (1973) (in Russian)

9. S.A. Arzhakov, N.F. Bakeev and V.A. Kabanov, Vysokomol. Soedin., A, 15, 1154 (1973) (in Russian)

10. S.A. Arzhakov, Doctorate (Chem.) Dissertation, Moscow: Karpov Institute of Physical Chemistry, 1975 (in Russian).

11. M.S. Arzhakov, S.A. Arzhakov, M.B. Lachinov, E.V. Nechaeva, and D.V. Trachenko, Intern. J. Polymeric Mater., (1996) (in press).

12. M.S. Arzhakov, S.A. Arzhakov, I.B. Kevdina and V.P. Shantarovich, in: Fizikokhimiya polimerov (Physics and Chemistry of Polymers), Tver', 2, 59 (1996) (in Russian).

13. M.S. Arzhakov, S.A. Arzhakov, V.V. Gustov, I.B. Kevdina and V.P. Shantarovich, Intern. J. Polymeric Mater., (1996) (in press).

14. M.S. Arzhakov and S.A. Arzhakov, Intern. J. Polymeric Mater., (1996) (in press).

15. K. Neki and P.H. Geil, J. Macromol. Sci., B, 8, 295 (1973).

16. H.E. Bair, M. Matsuo, W.A. Salmon and T.K. Kwei, Macromolecules, 5, 114 (1972).

17. J.L. Racich and J.A. Koutsky, J. Appl. Polym. Sci., 20, 2111 (1976).

18. M.Atsuta and D.T. Turner, J. Mater. Sci., Lett., 1, 167 (1982).

19. G.S.J. Yeh, IUPAC Macro 26th Symp., Maintz, Prepr. Short Commun., 2, 1176 (1979).

20. F.N. Kelley and D.R. Trainor, Polym. Bull., 7, 369 (1982).

21. P.H. Geil, Proc. 17th Intern. Symp., New York, London, 83 (1987).

22. D.A. Topchiev, Vysokomol. Soedin., A, 32, 2243 (1990) (in Russian).

23. J. Wendorff, Polymer, 23, 543 (1982).

24. V.P. Shantarovich, M.S. Arzhakov and I.B. Kevdina, Polym. Sci., B, 37, 171 (1995).

25. N. Tanio, Y. Koike and Y. Ohtsuka, Polym. J., 21, 259 (1989).

26. G.S.J. Yeh, J. Macromol. Sci., B, 6, 465 (1972).

27. I.J. Klement and P.H. Geil, J. Macromol. Sci., B, 5, 505 (1971).

28. Yu.A. Mikheev and L.N. Guseva, Khim. Fiz., 10, 724 (1991) (in Russian).

29. Yu.A. Mikheev, Intern. J. Polymeric Mater., 16, 221 (1992).

30. Yu.I. Matveev and A.A. Askadskii, Vysokomol. Soedin., A, 28, 1365 (1986) (in Russian).

31. G.V. Kozlov, D.S. Sanditov and V.D. Serdyuk, Polym. Sci., A, 35, 1742 (1993).

32. V.N. Belousov, G.V. Kozlov, A.K. Mikitaev and Yu.S. Lipatov, Dokl. Akad. Nauk SSSR, 313, 630 (1990) (in Russian).

33. I.I. Perepechko and A.V. Prokazov, Mekhanika polimerov, 749 (1977) (in Russian).

34. V.I. Surovtsev, S.S. Pelishenko, N.M. Zelenskaya-Surovtseva, A.A. Savitskaya and V.A. Degtyarev, Vysokomol. Soedin., A, 31, 339 (1989) (in Russian).

35. W.J. Koros and D.R. Paul, J. Polym. Sci., B, 16, 1947 (1978).

36. S. Subramanian, J.C. Heydweiller and S.A. Stern, J. Polym. Sci., B, 27, 1209 (1989).

37. W.R. Veith, J.M. Howell and H. Hsieh, J. Membr. Sci., 1, 177 (1976).

38. A.Ya. Polishchuk, L.A. Zimina, N.N. Madyuskin and G.E. Zaikov, Polym. Sci., A, 35, 80 (1993).

39. M. Lewin, H. Guttman and Y. Naor, J. Macromol. Sci., A, 25, 1367 (1988).

40. M.S. Arzhakov, Intern. J. Polymeric Mater., 28, 103 (1995).

41. V.N. Belousov, B.Kh. Kotsev and A.K. Mikitaev, Dokl. Akad. Nauk SSSR, 280, 1140 (1985) (in Russian).

42. I.I. Perepechko and O.V. Startsev, Akusticheskii Zhurnal, 22, 749 (1976) (in Russian).

43. I.I. Perepechko, Akusticheskie metody issledovaniya polimerov (Acoustic Studies in Polymers), Moscow: Khimiya, 1973 (in Russian).

44. V.A. Bershtein and V.M. Egorov, Differentsial'naya skaniruyushchaya kalorimetriya v fizikokhimii polimerov (Differential Scanning Calorimetry in Physical Chemistry of Polymers), Leningrad: Khimiya, 1990 (in Russian).

45. V.A. Bershtein, V.M. Egorov and V.A. Stepanov, Dokl. Akad. Nauk SSSR, 269, 627 (1983) (in Russian).

46. V.A. Bershtein, V.M. Egorov, Yu.A. Emel'yanov and V.A. Stepanov, Polym. Bull., 29, 98 (1983).

47. V.A. Bershtein and V.M. Egorov, Vysokomol. Soedin., A, 27, 2440 (1985) (in Russian).

48. G.V. Kozlov, V.N. Shogenov and A.K. Mikitaev, Dokl. Akad. Nauk SSSR, 298, 142 (1988) (in Russian).

49. Entsiklopediya polimerov (Encyclopedia of Polymers), Moscow: Sovetskaya Entsiklopediya, 1974 (in Russian).

50. P.V. Kozlov and S.P Papkov, Fizikokhimicheskie osnovy plastifikatsii polimerov (Physical and Chemical Foundations of Plasticization of Polymers), Moscow: Khimiya, 1982 (in Russian).

51. The Physics of Glassy Polymers, R.N. Haward, Ed., New York, Toronto: Wiley, 1973.

52. L.E. Nielsen, Mechanical Properties of Polymers and Composites, New York: Dekker, 1974.

53. V.I. Irzhak, B.A. Rozenberg and N.S. Enikolopyan, Setchatye polimery (Polymer Networks), Moscow: Nauka, 1979 (in Russian).

A New Approach for Preparation of Modern Polymeric Materials Based on Polymer Glasses

M.S. Arzhakov, S.A. Arzhakov and V.A. Chernavin

[*] Polymer Department, Faculty of Chemistry, Lomonosov Moscow State University, Vorob'evy Gory, Moscow, 119899 Russia

[**] Technological and Scientific Center of Organic Glasses, Dzerzhinsk, 606006 Russia

Temperature-induced relaxation of plastic deformation of bulk PMMA was studied. The coexistence of low-temperature and high-temperature components of relaxation was discussed in terms of structural inhomogeneity of polymer glasses. The problems concerning mechanical behavior of polymer materials and their modification were treated, in general, as the problems concerning controlled variations in the contributions from low-temperature and high-temperature components to temperature-induced relaxation of plastic deformation. The speculations on the role of structural inhomogeneity in the development of mechanical properties of polymer glasses allowed development of scientific and technological approach to prepare orientation-strengthened and shrinkproof polymeric materials.

Key words: polymer glasses, plastic deformation, relaxation, structure, shrinkproof materials

INTRODUCTION

Polymer glasses are widely used as modern materials (for example, engineering plastics) because of their unique properties, in particular, lowered brittleness and well-pronounced ductility as compared with low-molecular-mass glassy materials. For preparation of advanced polymeric materials and controlled modification of their physical and mechanical behavior one should understand the origin of these properties.

Studying temperature-induced relaxation of residual plastic deformation of polymer glasses provides an important information concerning physical and mechanical behavior of these materials. Temperature-induced relaxation of plastic deformation of polymer glasses shows a well-pronounced two-stage character, and involves low-temperature component at temperatures well below glass transition temperature T_g and high-temperature component at temperatures close to T_g [1 - 4]. Two-stage character of temperature-induced relaxation of plastic deformation allowed us to advance some speculations concerning the structural pattern of plastic deformation [5 - 7]. From this standpoint, plastic deformation and, in general, physical and mechanical behavior of polymer glasses is likely to be controlled by gradual involvement of segmental mobility within structural sublevels with different packing densities in deformation and relaxation processes.

On the other hand, temperature-induced relaxation of plastic deformation should be taken into account to predict behavior of polymeric materials under operating conditions.

A well-pronounced plastic deformation of polymer glasses

allows processability of polymer materials (rolling, stamping, etc.) and substantial improvement in their mechanical properties. One of the approaches to improve mechanical properties of commercial glassy polymers, in particular, toughness, is provided by the method of orientation strengthening.

However, such plastically deformed or oriented polymer materials and articles have one main disadvantage: as temperature exceeds polymer softening temperature, they appear to lose their shape and orientation, and, as a result, their mechanical properties dramatically deteriorate and approach those characteristic of undeformed and unoriented material. These processes are associated with high-temperature component of temperature-induced relaxation of plastic deformation or temperature-induced shrinkage of oriented glassy polymer. Moreover, the shrinkage of oriented polymer glasses takes place also in glassy state at temperatures well below glass transition temperature T_g via low-temperature component of relaxation.

In this work, we outlined the origin of low-temperature and high-temperature components of temperature-induced relaxation of plastic deformation and their contributions to deformation and relaxation processes in polymer glasses. The aim of the present paper was the development of scientific approach to control physical and mechanical behavior of materials based on glassy polymers and to prepare polymeric materials with specified properties.

EXPERIMENTAL

In this work, PMMA prepared by bulk polymerization (M_η = $3x10^6$) was used. Dibutyl phtalate (DBPh) was used as plasticizing agent. The content of plasticizing agent was varied from 1 to 10 mass%. The copolymers of MMA with bifunctional monomer ethylene glycol dimethacrylate (EGDM) were used as cross-linked PMMA samples. The development of network took place immediately during bulk polymerization. The content of EGDM in polymerization system was 1.0, 5.2, and 11.2 mol%.

The test samples were cylinders with a height of 30 mm and a diameter of 15 mm. Prior to testing, polymer samples were annealed at temperatures, which exceeded glass transition temperature by 10 - 15 K. Then, they were slowly cooled down to room temperature.

The samples of initial and plasticized PMMA were uniaxially compressed at temperatures 293, 320, 333, 353, and 383K with a strain rate of 0.1 mm/min using an UTS-100 tensile machine (Germany). After that, they were cooled with liquid nitrogen in a stressed state. To study temperature-induced relaxation of residual deformation, the free-standing samples were heated from 213K with the rate of heating 1 K/min. Temperature-induced relaxation of residual deformation was estimated by measuring the height of the samples with an accuracy of ± 0.001 mm.

The samples of the cross-linked PMMA were uniaxially compressed at 293 K with a strain rate of 0.1 mm/min. Then, they were unloaded with the same rate at this temperature. Temperature-induced relaxation of residual deformation was studied starting with 293 K.

Orientation-strengthened samples were prepared by multiaxial plane drawing of PMMA sheets to strains of 60 and 120% at temperatures, which exceeded glass transition temperature by 30 K. Then, the stressed samples were slowly cooled down to room temperature.

To determine ultimate strain of initial and orientation-strengthened PMMA samples, the dumbbell samples were tested under uniaxial drawing at a strain rate of 10 %/min at 293K. Polymer toughness was determined according to Izod [8] at 293K. Shrinkage of orientation-strengthened PMMA samples was estimated by changes in their linear dimensions on heating to temperature, which exceeded softening temperature by 20 K.

RESULTS AND DISCUSSION

Figure 1 shows a typical temperature dependence of relaxation of residual deformation ε_{res} in PMMA samples. Temperature-induced relaxation of residual deformation takes place within two well-defined temperature regions and involves low-temperature ε_1 and high-temperature ε_2 components. This experimental evidence suggests that on heating the relaxation of plastically deformed samples is related to restructuring of, at least, two different structural elements.

The ratio between the components associated with low-temperature and high-temperature relaxation was studied as a function of strain. For the samples compressed to rather low strain, the complete temperature-induced relaxation of residual deformation takes place at temperatures well below glass transition temperature T_g (Fig. 2). On the contrary, for the samples compressed to rather high strains, a complete relaxation is observed on heating up to temperatures close to T_g. Note that, in this case, the low-temperature relaxation is also involved.

Figure 3 presents low-temperature ε_1 and high-temperature ε_2 components of temperature-induced relaxation of residual deformation as a function of strain and the stress-strain diagram corresponding to uniaxial compression of PMMA samples. This figure shows that, for the samples with strains below yield strain ε_y, relaxation of residual deformation proceeds via low-temperature mode ε_1. With increasing strain above ε_y, the appearance and growth in high-temperature component ε_2 at constant ε_1 is observed.

Two-stage character of temperature-induced relaxation of residual deformation allows a convincing conclusion that plastic deformation of polymers involves two modes which relax via low-temperature and high-temperature components. In other words, this implies realization of two modes of molecular mechanism of both plastic deformation and its relaxation in polymers.

As this takes place, elastic relaxation of plastic deformation via high-temperature component ε_2 is observed in a narrow temperature region in the vicinity of T_g. This relaxation is associated with transition of excited chain conformations to initial equilibrium state via segmental motion within the polymer bulk.

Alternatively, elastic relaxation of plastic deformation via low-temperature component ε_1 is observed at temperatures well below T_g in a wide temperature range within polymer glassy state. Note, that each temperature is associated with a certain value of relaxation of this fraction of plastic deformation. In other words, we deal with a temperature-dependent mode of elastic recovery of plastic deformation.

At the present time, the two-stage character of temperature-induced relaxation of polymer plastic deformation is treated in terms of two standpoints: step by step recovery of small-scale plastic shear transformations and excited chain conformations [3, 4] (i) and structural inhomogeneity of glassy polymers [5 - 7] (ii).

Within the framework of the latter approach, glassy polymer is characterized by a set of structural sublevels with different packing densities and degrees of ordering.

During recent decades, the problems related to structure of glassy polymers are being widely discussed in literature. Let us mention some of them.

As demonstrated by electron microscopy (for example, [9 - 11]), positron annihilation [7, 12], light scattering measurements [13], small-angle X-ray scattering [14], and wide-angle X-ray scattering [15], polymer glasses are characterized by structural inhomogeneity related to the existence of heterogeneous ordered regions with higher packing densities (domains, clusters) with dimensions up to several nanometers. These domains are joned with each other by numerous tie chains, which constitute the regions with lower packing densities and ordering.

Experimental evidence obtained provided a fundamental basis for the development of various structural models of glassy polymers. The difference between the models advanced is related to the type of principal morphological structural unit: fibril [16], domain [17, 18], globule [19 - 21], and cluster [22, 23]. However, at the present time, a certain lack of convincing experimental evidence does not allow one to decide in favor of this or that particular model. Hence, in these work, our speculations will be based on only unequivocal conclusion concerning structural inhomogeneity of glassy polymer, which is associated with the existence of local structural regions with different packing of macromolecules or their fragments.

Each of these sublevels is associated with its own relaxation time and activation energy of segmental mobility. Molecular mechanism of the low-temperature relaxation is similar to that of high-temperature mode. According to this mechanism, each temperature is related to the appearance of segmental mobility within a certain structural sublevel. Let us emphasize that a possible realization of segmental mobility in local structural sublevels at temperatures below T_g is discussed in [24 - 27]. In addition, the appearance of local segmental mobility at temperatures below T_g is likely to be controlled by accumulation of inner stresses [28] as well as by asymmetry of the field of intermolecular interaction provided by free volume microregions [29]. As a result, a decrease in effective energy of activation of segmental mobility is observed, and relaxation of segments in local structural sublevels is allowed within glassy state.

Appearance of low-temperature and high-temperature components of relaxation of plastic deformation allowed us [5 - 7] to advance the structural pattern of plastic deformation of glassy polymers concerning a gradual involvement of various structural elements with different packing densities into deformation and relaxation. From this standpoint, the ratio between low-temperature component ε_1 and high-temperature component ε_2 (or contributions of ε_1 and ε_2 to temperature-induced relaxation of residual deformation) describes the development of plastic deformation of polymer glasses. Let us consider in detail the role of the above components of relaxation in the development of physical and mechanical behavior of polymer glasses.

Low-temperature component is likely to be responsible for low-temperature plasticity of glassy polymers and high reversible strains, which are known to be valuable properties of polymer plastics. This important feature of deformation and relaxation behavior of polymer glasses seems to be intimately related to their unique properties, in particular, lowered brittleness as compared with low-molecular-mass glasses. Let us emphasize that the strength properties of polymer, such as elastic modulus and yield stress, and the contribution from low-temperature component to relaxation of residual deformation are well correlated [5].

On the other hand, with increasing operating temperature, low-temperature and high-temperature components of temperature-induced relaxation of plastic deformation are the main reason of poor shape stability of deformed polymer articles and shrinkage of oriented materials. As a result, a dramatic deterioration in mechanical properties of materials is observed.

In connection with this, the problems concerning preparation of polymeric materials with specified properties and controlled mechanical behavior under operating conditions should be discussed in two aspects.

At first, for oriented and unoriented polymeric materials at a given operating temperature, the problems concerning modification of their physical and mechanical behavior should be treated, in general, as the problems concerning controlled variation in the ratio between low-temperature and high-temperature components of temperature-induced relaxation of residual deformation.

Secondly, for oriented polymeric materials in a wide range of operating temperatures, lowering or suppression of both low-temperature and high-temperature components of relaxation presents a challenging task for many scientists and engineers. A happy solution of this problem offers numerous advantages. If we manage to prevent low-temperature relaxation mode, this implies a substantial broadening of temperature interval of operating conditions up to T_g. But if both modes of relaxation are prevented, this will allow much progress in development of a new generation of orientation-strengthened and shrinkproof polymer materials based on well-known commercial polymers.

To outline scientific approaches to control the ratio between low-temperature and high-temperature components of relaxation, let us consider the temperature-induced relaxation of plastic deformation for various PMMA samples.

Figure 4 shows temperature dependences of residual deformation for PMMA samples plasticized with DBPh. For the lack of space, relaxation curves are presented only for the samples compressed to strains of 14 and 20 %. Plasticization of PMMA leads to appearance and growth in high-temperature component of relaxation and decrease in low-temperature component. In contrast, the cross-linking of PMMA samples with EGDM is associated with increasing low-temperature component of relaxation and degeneration of high-temperature component (Fig. 5). As a result, at relatively high content of EGDM (5 - 11 mol %), relaxation of plastic deformation proceeds via low-temperature component at temperatures well below T_g. These experimental evidence suggests possibilities, which allow one to control contributions from low-temperature and high-temperature components to relaxation of residual deformation via physicochemical modification.

For PMMA samples compressed at different temperatures, experimental evidence suggests that preventing low-temperature component ε_1 is provided by increasing deformation temperature T_{def} (Fig. 6). As the temperature of deformation increases, a contribution from the low-temperature component ε_1 to temperature-induced relaxation of residual deformation decreases. When compression of polymer samples is carried out at temperatures in the vicinity of T_g, relaxation of component ε_1 is likely to take place during the deformation. In this case, the total relaxation is primarily controlled by high-temperature component ε_2 Fig. 6, curves 5).

This approach may be served as a basis for traditional technology of preparation of orientation-strengthened polymer glasses. This technology involves orientation (tensile drawing) of a polymer samples above glass transition temperature and their further cooling in a stressed state. Materials and articles prepared via this method are characterized by a stability of their both shape and orientation and, hence, resulting mechanical properties up to temperatures not exceeding T_g.

The consideration of the plastic deformation and its further relaxation in polymer glasses from the standpoint of the structural inhomogeneity of these materials allowed us to solve the problem concerning suppression of high-temperature component of relaxation of plastic deformation via a new approach to orientation strengthening of polymers. Technological application of this approach makes it possible to prepare orientation-strengthened shrinkproof polymer materials based on PMMA. Mechanical characteristics of these materials appear to be comparable to those of a conventional oriented glass produced by a traditional technological method. However, the main advantage of such materials is associated with their high thermal stability, i.e., retention of shape, orientation and, finally, mechanical characteristics in a wide temperature range and even after heating the samples up to temperatures above the softening temperature of PMMA (393 K). The comparison of the properties of PMMA samples prepared via the new technology (PMMA-I and PMMA-II) with those of unoriented PMMA and PMMA prepared via traditional technology is presented in Table 1.

This new approach of orientation strengthening of polymer materials can be effectively used in many areas of technological production of the finished articles based on various polymers in the course of extrusion or fiber spinning from melts and solutions. Furthermore, this new method allows polymer processing just in the course of polymerization.

CONCLUSION

The coexistence of low-temperature and high-temperature components of temperature-induced relaxation of residual plastic deformation is considered in terms of structural inhomogeneity of polymer glasses. From this standpoint, the development of physical and mechanical behavior of glassy polymers can be presented as gradual involvement of structural sublevels with different packing densities to deformation and relaxation. At macroscopic level, this process manifests itself as a temperature-dependent mode of elastic recovery, which is specific feature of glassy polymers.

Low-temperature component is likely to be responsible for low-temperature plasticity of glassy polymers and high reversible strains, which are known to be valuable properties of polymer plastics. This important feature of deformation and relaxation behavior of polymer glasses is intimately related to a number of their unique properties, in particular, lowered brittleness as compared with low-molecular-mass glasses. On the other hand, this relaxation mode was shown to be responsible for low-temperature shrinkage (loss in orientation) of oriented polymer materials on heating the samples at temperatures below T_g.

Hence, for polymer materials based on glassy polymers mechanical behavior is mainly controlled by low-temperature component of temperature-induced relaxation of plastic deformation of polymer glasses. In connection with this, the problems concerning preparation of modern polymeric materials based on glassy polymers should be treated, in general, as the problems concerning controlled variations in the ratio between low-temperature and high-temperature components of relaxation. The experimental evidence obtained suggests possibilities, which allow one to control variations in the contribution from low-temperature component to relaxation via physicochemical modification.

The speculations concerning the role of structural inhomogeneity in the development of both mechanical and relaxation properties of polymer glasses make it possible to put forth new technology of preparation of orientation-strengthened and shrinkproof polymer materials.

Table 1

MECHANICAL CHARACTERISTICS OF PMMA GLASSES AS OBTAINED BY NEW TECHNOLOGY

Material		Strain %	Ultimate strain, %	Toughness, kJ/m^2	Shrinkage on heating above softening temperature, %
Unoriented PMMA		0	6	12	0
PMMA Orientational Strengthening by Traditional Technology		60	26	25	100
		120	6	55	100
PMMA Orientational Strengthening by New Technology					
	PMMA-I	60	16	22	52
		120	23	50	40
	PMMA-II	60	6	20	0
		120	6	26	0

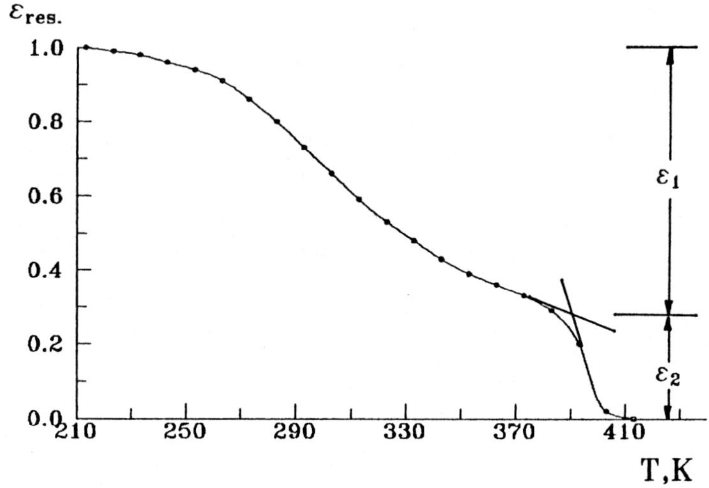

Fig. 1. Typical temperature dependence of relative residual deformation ε_{res} of PMMA. Strain: 20%. T_{def} = 293 K.

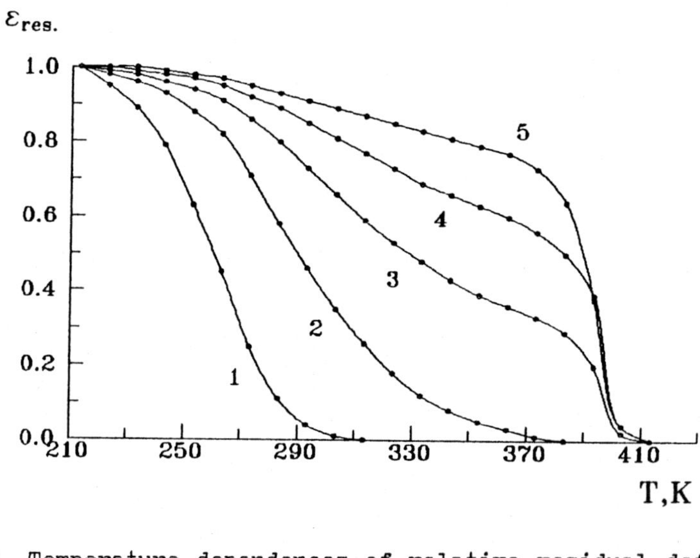

Fig. 2. Temperature dependences of relative residual deformation ε_{res} of PMMA samples. Strain: 6 (1), 14 (2), 21 (3), 30 (4), and 49% (5). T_{def} = 293 K.

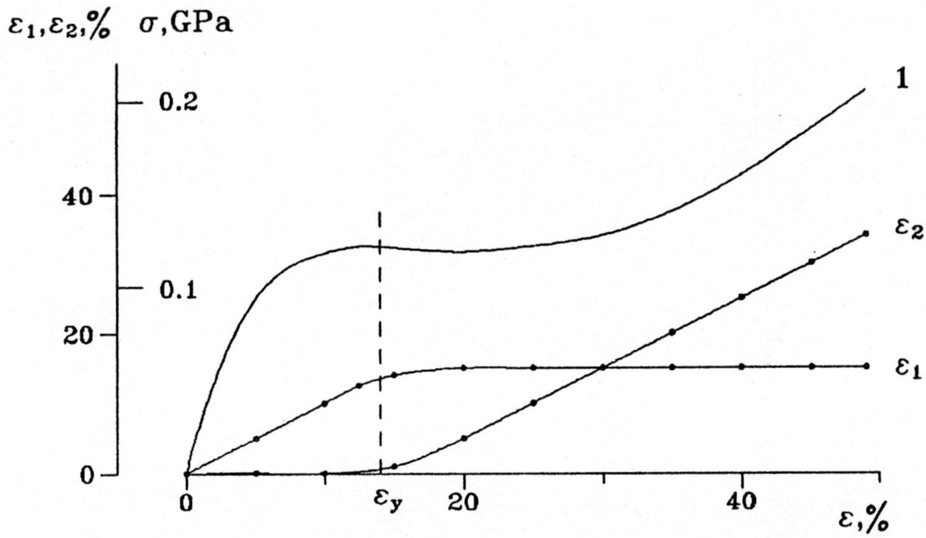

Fig. 3. Stress-strain curve corresponding to uniaxial compression of PMMA samples at 293 K (1) and low-temperature ε_1 and high-temperature component ε_2 of relaxation of residual deformation versus total strain ε.

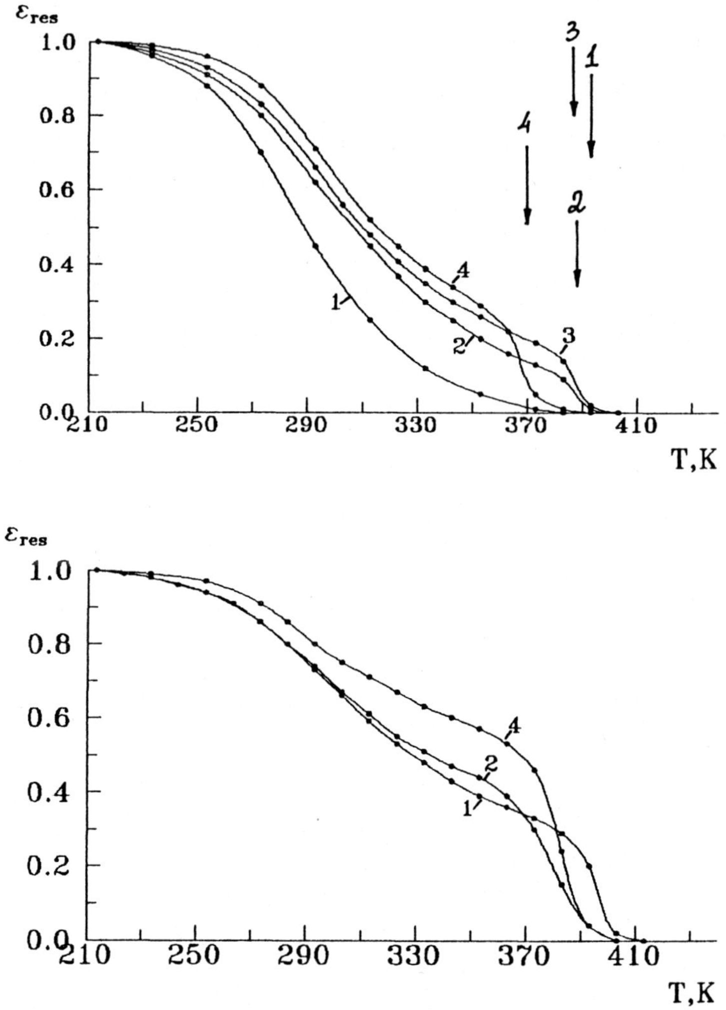

Fig. 4. Temperature dependences of relative residual deformation ε_{res} of plasticized PMMA samples. Content of plasticizing agent DBPh: 0 (1), 1 (2), 5 (3), and 10 mass % (4). Strain: 14 (A) and 20 % (B). T_{def} = 293 K. The arrows show T_g of the corresponding polymer samples.

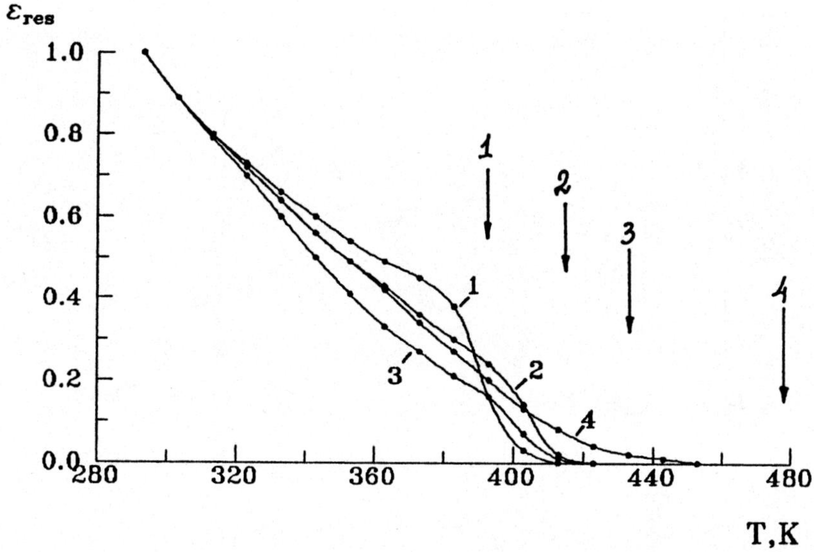

Fig. 5. Temperature dependences of relative residual deformation ε_{res} of cross-linked PMMA samples. Content of cross-linking agent EGDM: 0 (1), 1 (2), 5.2 (3), and 11.2 mol. % (4). Strain: 20 %. T_{def} = 293 K. The arrows show T_g of the corresponding polymer samples.

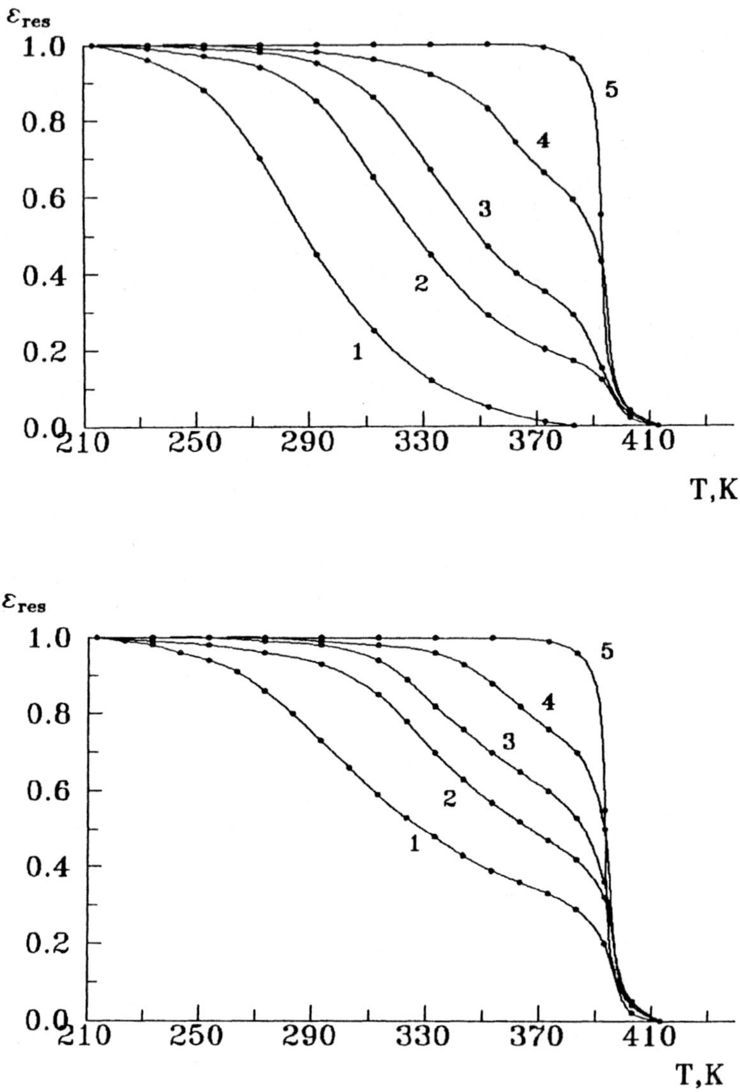

Fig. 6. Temperature dependences of relative residual deformation ε_{res} of PMMA samples. T_{def} = 293 (1), 320 (2), 333 (3), 353 (4), and 383K (5). Strain: 14 (A) and 20 % (B).

REFERENCES

1. S.A. Arzhakov and V.A. Kabanov, Vysokomol. Soedin., B, 13, 318 (1971) (in Russian)

2. A.E. Skorobogatova, S.A. Arzhakov, N.F. Bakeev and V.A. Kabanov, Dokl. Akad. Nauk SSSR, 211, 151 (1973) (in Russian)

3. E. Oleynik, Progr. Colloid & Polym. Sci., 80, 140 (1989)

4. E.F. Oleinik, O.B. Salamatina, S.N. Rudnev and S.V. Shenogin, Polymer Science, 35, 1532 (1993)

5. M.S. Arzhakov and S.A. Arzhakov, Intern. J. Polymeric Mater., 29, 249 (1995).

6. M.S. Arzhakov and S.A. Arzhakov, in: Fizikokhimiya polimerov (Physics and Chemistry of Polymers), Tver', 2, 64 (1996) (in Russian).

7. M.S. Arzhakov, S.A. Arzhakov, I.B. Kevdina and V.P. Shantarovich, in: Fizikokhimiya polimerov (Physics and Chemistry of Polymers), Tver', 2, 59 (1996) (in Russian).

8. S. Turner, Mechanical Testing of Plastics, London: Iliffe Books, (1973)

9. M. Atsuta and D.T. Turner, J. Mater. Sci., Lett., 1, 167 (1982).

10. F.N. Kelley and D.R. Trainor, Polym. Bull., 7, 369 (1982).

11. D.A. Topchiev, Vysokomol. Soedin., A, 32, 2243 (1990) (in Russian).

12. V.P. Shantarovich, M.S. Arzhakov and I.B. Kevdina, Polymer Science, B, 37, 171 (1995).

13. N. Tanio, Y. Koike and Y. Ohtsuka, Polym. J., 21, 259 (1989).

14. G.S.J. Yah, IUPAC Macro 26th Symp., Mainz, Prepr. Short Commun., 2, 1176 (1979).

15. J. Wendorff, Polymer, 23, 543 (1982).

16. S.A. Arzhakov, N.F. Bakeev and V.A. Kabanov, Vysokomol. Soedin., A, 15, 1154 (1973) (in Russian)

17. G.S.J. Yeh, J. Macromol. Sci., B, 6, 465 (1972)

18. I.J. Klement and P.H. Geil, J. Macromol. Sci., B, 5, 505 (1971)

19. Yu.A. Mikheev and L.N. Guseva, Khim. Fiz., 10, 724 (1991) (in Russian).

20. Yu.A. Mikheev, Intern. J. Polymeric Mater., 16, 221 (1992).

21. Yu.I. Matveev and A.A. Askadskii, Vysokomol. Soedin., A, 28, 1365 (1986) (in Russian).

22. V.N. Belousov, G.V. Kozlov, A.K. Mikitaev and Yu.S. Lipatov, Dokl. Akad. Nauk SSSR, 313, 630 (1990) (in Russian)

23. G.V. Kozlov, D.S. Sanditov and V.D. Serdyuk, Polymer Science, A, 35, 1742 (1993).

24. V.A. Bershtein and V.M. Egorov, Differentsial'naya skaniruyushchaya kalorimetriya v fizikokhimii polimerov (Differential Scanning Calorimetry in Physical Chemistry of Polymers), Leningrad: Khimiya, 1990 (in Russian).

25. V.I. Surovtsev, S.S. Pelishenko, N.M. Zelenskaya-Surovtseva, A.A. Savitskaya and V.A. Degtyarev, Vysokomol. Soedin., A, 31, 339 (1989) (in Russian).

26. V.N. Belousov, B.Kh. Kotsev and A.K. Mikitaev, Dokl. Akad. Nauk SSSR, 280, 1140 (1985) (in Russian).

27. I.I. Perepechko, Akusticheskie metody issledovaniya polimerov (Acoustic Studies in Polymers), Moscow: Khimiya, 1973 (in Russian).

28. M.S. Arzhakov and S.A. Arzhakov, Intern. J. Polymeric Mater., (1996) (in press).

29. G.V. Kozlov, V.N. Shogenov and A.K. Mikitaev, Dokl. Akad. Nauk SSSR, 298, 142 (1988) (in Russian).

Structural Aspects of Modification of Polymers with Low-Molecular-Mass Substances

M.S. Arzhakov

Polymer Department, Faculty of Chemistry, Moscow State University, Vorob'evy Gory, Moscow, 119899 Russia

Specific features of polymer modification based on complexation between low-molecular-mass modifying agents and polymers were discussed in terms of structural inhomogeneity of amorphous glassy polymers. Complexation between low-molecular-mass compounds (LMs) and polymer macrochains in structural regions with different packing densities was shown to be responsible for the distribution of LMs between two forms, which are characterized by different energetics of interaction with polymer matrix. The occurrence of such distribution was controlled by chemical structure of LM molecules and local packing of polymer macrochains. This distribution was responsible for a complicated character of molecular dynamics in the polymer/LM systems and mechanical behavior of such materials in whole. The cases, when macroscopic mechanical behavior of as-modified polymers was primarily controlled by the form, in which the modifying agent existed in polymer matrix, were considered. The future development of this direction of polymer modification was discussed and possibilities of its practical application for controlled improvement in mechanical response of polymer materials were outlined.

Key words: polymers, low-molecular-mass compounds, interaction, complexation, structure, physical and mechanical properties

INTRODUCTION

In many cases physical modification of polymer materials may be achieved by introduction of compatible low-molecular-mass additives (plasticizing agents, dyes, reinforcing agents, flame-retardants, etc.) into polymers. Compatibility of the components is controlled by nonchemical interaction between molecules of low-molecular-mass compounds (LMs) and polymer chains. Such interaction is provided by hydrogen bonding, electrostatic, dispersion, and similar interactions, which, in general case, are responsible for complexation between the components of polymer composites. Physical and mechanical characteristics of the as-modified polymer materials are decided, within the first approximation, by energetics of interaction between polymer chains and molecules of modifying agent.

In some cases, LM molecules are known to be distributed within polymer matrix between two different forms, which are characterized by different energy of interaction with polymer [1-4]. As a result, physical and mechanical behavior of such complex polymer/LM systems shows a number of peculiar features. In particular, one and the same modifying agent may have quite an opposite effect on mechanical properties of the polymer depending on the form, in which it exists in polymer matrix [3, 5].

In connection with this, the studies on distribution of LM in polymer matrix between two forms are of great interest and will allow one to gain, on one hand, a deeper insight into the mechanism of this phenomenon and, on the other hand, to take advantages of this phenomenon for controlled modification of the polymer materials.

The aim of the present work is to summarize the published data on interaction of LMs with polymers. The principle idea of this work is not to cover all the related aspects but to attract the attention of scientists to this fascinating problem and to outline general ways of its further development.

STRUCTURAL FEATURES OF POLYMER INTERACTION WITH LOW-MOLECULAR-MASS COMPOUNDS

In some cases, introduction of low-molecular-mass compounds (LMs) into the polymer materials is accompanied by distribution of LM between two different forms, which are characterized by different energy of interaction with polymer matrix. This phenomenon was described in publications devoted to modification of PA-6 with oxyaromatic compounds (OACs) [1, 2], sorption of Br_2 by PAN fibers from aqueous bromine solutions [3], transport of dioxydine in ternary system based on dioxydine, water and copolymer of N-vinylpyrrolidone with methyl methacrylate [4].

In all cases, characteristic features of this phenomenon are the following. LMs, which were introduced into the polymer matrix from the swelling solutions (in all cases studied, aqueous solutions were used), appear to be distributed within the polymer matrix between two different forms, which are characterized by reversible or irreversible binding with polymer matrix. Hereinafter, such forms are referred to as reversibly (RB) and irreversibly bound (IB) states. As compared with IB form, RB form may be easily removed from the polymer matrix by keeping the polymer samples in a pure solvent (water).

The common feature of all the above polymer systems is related to the fact that LMs show a high affinity to polymer matrices. Thermodynamical studies of sorption of Br_2 by PAN fibers [3] and OAC by PA-6 films [2] revealed that partition coefficients of LMs between polymer matrices and aqueous media are rather high, that is, preferential sorption of LMs from their aqueous solutions is observed. In the cases studied, the origin of high affinity between polymers and LMs is related to complexation of molecules of LMs and polymer chains. In particular, interaction between Br_2 and PAN is associated with charge-transfer complexation between Br_2 and nitrile polymer groups [3]. In the case of the PA-6/OAC systems, complexation between polymer chains and molecules of LM is provided by hydrogen bonding of phenol hydroxyls of OAC with amide polymer groups [1, 6-11].

Kinetic and thermodynamic studies of sorption of Br_2 by PAN fibers from the bromine aqueous solutions at temperatures from 25 to 40°C allowed Lewin *et al.* [3] to advance the following mechanism of distribution of Br_2 in PAN matrix between IB and RB forms.

Interaction between PAN and LM was interpreted in terms of structural heterogeneity of amorphous component of semicrystalline polymer, that is, the coexistence of, at least, two different noncrystalline phases with different packing density and ordering. One should mention that the structural heterogeneity of the amorphous regions in semicrystalline polymers and completely amorphous polymers is an intriguing problem, which is being vigorously discussed in literature [12-18].

According to this approach, sorption of Br_2 by PAN fibers involves two stages [3]. The first stage is associated with polymer swelling in aqueous bromine solution. As a result, bromine appears to be dissolved in loosely packed regions of polymer amorphous phase and RB form is developed. At the second stage, bromine molecules diffuse to more densely packed regions, which are unaccessible to water molecules. In this case, bromine molecules disrupt a part of interchange dipole interactions of CN groups, produce relatively stable charge-transfer complexes with nitrile groups, and become immobilized, forming thus IB portion of LM.

To conclude, according to Lewin [3] the development of IB from of LM in polymer matrix is accounted for by diffusion of Br_2 molecules into the densely packed domains and their subsequent immobilization via charge-transfer complexation. In this case, formation of the charge-transfer complexes in the loosely packed regions is likely to be absent.

Thermodynamic and kinetic studies of sorption of synthetic tanning agent (BNS) by PA-6 films at 100°C showed [2] that the mechanism advanced in [3] may also be invoked to explain the distribution of LM in polymer matrix between two IB and RB forms.

However, the question arises: Why does immobilization of the molecules of LMs in the polymer matrix via complexation (if this very factor is responsible for irreversible binding of LM in polymer matrix) proceed only in densely packed regions of the amorphous phase? This question seems to be of vital importance for the PA-6/OAC systems because phenol hydroxyls of OACs show very high tendency for complexation with amide groups [6, 11]. In connection with this, one can hardly explain the absence of interaction between polymer chains and OACs molecules in the loosely packed regions of amorphous phase of PA-6.

To answer the above question, a comprehensive study of mechanism of distribution of OACs molecules between RB and IB forms in the amorphous regions of PA-6 based on speculations concerning the complexation between the components involved was carried out [19]. Resorcinol (I), dioxydiphenylsulfone (II), synthetic tanning agent BNS (III), and resorcinol-formaldehyde novolac (IV) were used as OACs.

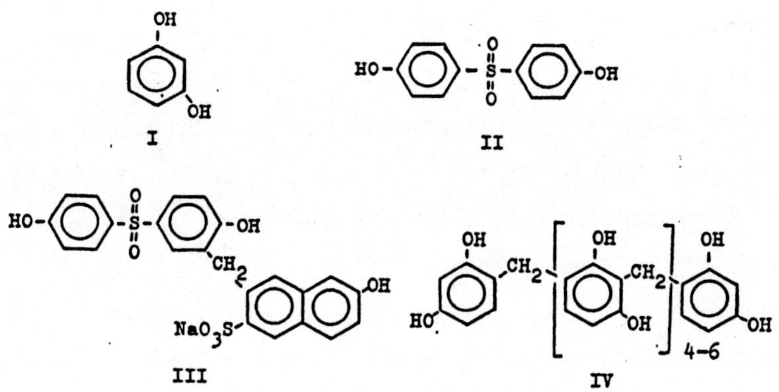

I, II, III, IV

As was shown, only compounds III and IV are able to produce IB form, and compounds I and II are not. The studies on sorption of water vapors by the PA-6/OAC samples at 20°C allowed one [2] to conclude that IB form of compound III is primarily localized in the densely packed regions of the amorphous phase of PA-6, which are unaccessible for water molecules at sorption temperatures in a whole range of relative humidities. The RB form occupies disordered regions of amorphous phase, the water penetration into which is known to control the water sorption by the polymer samples [1, 20].

In the case of compound IV, the formation of IB form takes place in the loosely packed regions of the amorphous phase of PA-6. However, there is still an open question: Do molecules of compound IV penetrate the densely packed domains of amorphous phase in PA-6 or they are localized only in the loosely packed regions?

From the standpoint of complexation, the distribution of LM between two forms may be represented as a result of occurrence of two reactions of complexation between the components of the PA-6/OAC system:

$$[OAC] + [CONH] \leftrightarrow [OAC \cdots CONH] \quad (1)$$

$$[OAC] + [CONH] \rightarrow [OAC \cdots CONH] \quad (2).$$

The difference between the above reactions is related to reversible character of reaction (1) and irreversible character of reaction (2). The subsequent treatment of the PA-6/OAC samples with a pure solvent (water) is associated with shifting the equilibrium state of reaction (1) to the left until complete dissociation of the complex. At the same time, the complex formed via reaction (2) shows no dissociation under the above conditions.

Examination of published evidence on complexation of PET [21] and poly(ethylene oxide) (PEO) [22] with resorcinol in melt, poly(N,N-dimethylacrylamide) with phenol-formaldehyde resins in the solutions [23] and the general speculations concerning the mechanism of complexation between polymer and oligomer chains [24] allowed one to specify two interdependent conditions for the development of stable intermolecular complexes:

1. Realization of joint dense packing of the molecules of the reacting compounds [21, 22]. In other words, reacting components should be complementary: geometric structure of the interacting species should not prevent the development of intermolecular bonds [24];

2. The complexation reaction should involve a certain sequence of active complexing groups, the number of which is equal to or exceeds critical one [23, 25]. This condition implies cooperativity of a set of intermolecular bonds in the complexes with polymer chains. Just cooperativity of interaction is primarily responsible for relatively high stability of the reaction products as compared with that of the complexes between the low-molecular-mass analogs.

Let us first discuss what do different authors mean when speaking about a "stable complex".

For example in [24], "stable complex" is treated from the kinetic standpoint. In the case of complexation reactions between polymer chains and oligomers, an increased stability of the formed complex was accounted for by an increase in the constant of equilibrium, i.e., by shifting the equilibrium towards the complex formation. The above reactions were shown to possess reversible character. In particular, the reaction of complexation between polymer and oligomer chains is reversible under varying the concentration of oligomer in the solution: a decrease in concentration of the oligomer is accompanied by shifting the equilibrium towards accumulation of the initial reagents until complete dissociation of the complex.

On the other hand, in the case of the complexation between poly(N,N-dimethylacrylamide) and phenol-formaldehyde resins [23] in organic solvents, the stable complex was referred to as the reaction product, which precipitated under the reaction conditions. Let us emphasize that the formed complex showed no dissociation under washing with a pure solvent. This evidence suggested irreversible character of the complexation reaction.

The similar situation was observed in the case of the complexation between aliphatic PA with OACs in the solutions of formic acid [6, 26] and alcohol [26]. In this case, compound III and sulfonated phenol-formaldehyde oligomer were used as OACs. Interaction of the above compounds with polymers (in particular, with PA-6) is accompanied by precipitation of the complex, which shows dissociation neither in a pure solvent nor in a boiling water.

A certain inconsistency in treating the term of "stable complex" does not conflict with the above conditions for complexation, which were advanced to explain the mechanism of specific interaction of OACs with polymer chains in the amorphous regions of PA-6. However, to prevent any terminological misunderstanding, in this work "stable complex" is the complex, which shows no dissociation under treating in a pure solvent, i.e., which is formed via reaction (2).

The occurrence of reaction (2) in amorphous regions of PA-6 is allowed when the number of active complexing groups in OAC molecules is high enough, and such groups are able to react with polymer amide groups so that the number of reacting groups should be equal to or higher than critical one. The latter requirement is especially important for the complexation reactions in solid bodies, since as compared with solutions and melts, the macromolecular mobility in the solid phase is substantially inhibited (in our case, in amorphous regions of PA-6 swollen in aqueous solutions of OAC).

Inhibited mobility of polymer chains prevents the possibility for phenol hydroxyls of OACs and polymer amide groups to approach each other within the distance necessary for the development of hydrogen bonds. Furthermore, in the amorphous regions of PA-6, as compared with solution or melt, the mutual arrangement of macromolecules and, hence, the chain fragments with amide groups is fixed. Mutual arrangement of polymer amide groups is primarily controlled by polymer packing density in amorphous regions of PA-6. The heterogeneity in packing density in PA-6 was discussed in [18].

From this standpoint, the occurrence of reaction (2), i.e., the development of

IB form, in the amorphous phase of PA-6 takes place only in the local structural regions which serves as a specific "trap" for OAC molecules. Just within such a trap, the above conditions for the development of stable complex are fulfilled and irreversible binding of OAC with polymer matrix is promoted. Let us discuss the experimental evidence obtained for complexation of PA-6 with OACs from this standpoint.

The development of only RB form for compounds I and II suggests that the chemical structure of the molecules of the above compounds (molecular sizes and number of the complexing groups) prevents the fulfillment of the necessary condition for the formation of stable intermolecular complex via reaction (2). In this case, interaction of phenol hydroxyls with polymer amide groups is described by reaction (1) and reversible character of binding of LM with polymer is realized.

When passing to OAC III, increase in molecular sizes and number of complexing hydroxyl groups result in appearance of IB form of LM. This implies that the number of active complexing groups is sufficient to produce stable complex with PA-6. However, the formation of IB form of OAC III takes place only within the densely packed regions of the polymer amorphous phase. This experimental evidence allows one to conclude that macromolecular arrangement only in the densely packed regions provides the conditions of cooperative interaction of all complexing groups of OAC with amide groups of PA-6, occurrence of reaction (2), and irreversible binding of OAC molecules with polymer matrix. In other words, the chemical structure of compound III molecules is so that the "traps" for them exist only in the densely packed regions of amorphous phase of PA-6. In the case of more loosely packed regions, the mutual arrangement of molecules in such regions prevents simultaneous interaction of amide groups with phenol hydroxyls of compound III in quantity necessary for the occurrence of reaction (2). As a result, the interaction between the components of the PA-6/OAC system is described by reaction (1) and the formation of only RB form is observed.

With further increasing the molecular sizes and the number of the active complexing groups, i.e., compound IV, one can observe the development of IB form also (or only) in the loosely packed regions of the amorphous phase of PA-6. This evidence suggests that, as compared with molecules of compound III, the molecules with such molecular structure are able to find the "traps" in the regions with lowered packing density.

The above speculations allow us to generalize the mechanism of distribution of LMs between IR and RB forms as advanced in [3].

Sorption by polymer of LMs from their aqueous solutions involves two stages. The first stage is associated with diffusion of molecules of LMs into the swollen regions of the polymer amorphous phase, which are characterized by lowered packing density. The second stage is related to diffusion of molecules of LMs from loosely to densely packed regions. The sorption is accompanied by binding of molecules of LMs by polymer chains both in loosely and densely packed regions of the polymer amorphous phase via complexation between the components.

In general case, complexation between the molecules of LMs and polymer chains in polymer amorphous phase can be described by reactions (1) and (2). The formation of IB form of LMs is described by reaction (2). The occurrence of this reaction is provided by fulfillment of the above conditions of development of stable intermolecular complex. The possibility of fulfillment of these conditions in the polymer amorphous phase is decided by the chemical structure of molecules of LMs (molecular size and functionality) and packing density of polymer chains. In connection with this, IB form of LMs may be realized either in loosely or in densely packed regions of polymer amorphous phase or does not form at all.

However, the mechanism of diffusion of molecules of LMs into the densely packed regions and their withdrawal from such regions is still unclear. The penetration of molecules of LMs into the densely packed regions proceeds from their aqueous solutions, which are able to swell polymer matrix. On the other hand, desorption of molecules of LMs from the densely packed regions implies transition of molecules of LMs from polymer matrix into the a pure solvent - water. The question arises: Does water swell the above regions or, in other words, whether such regions are accessible to water molecules?

The data reported in [3] clearly demonstrated that the densely packed regions of amorphous phase of PAN are unaccessible for water molecules under sorption and desorption of Br_2 from aqueous solutions. The formation of IB form of dioxydine in copolymer of N-vinylpyrrolidone with methyl methacrylate was also treated [4] within the standpoint of the existence of structural zones in the amorphous regions of moderately hydrophilic polymers, which are unaccessible for water molecules. However, in the case of the PA-6/OAC systems such conclusions do not seem quite obvious.

As was shown in [2, 19], the densely packed regions of the amorphous phase of PA-6 are unaccessible for water molecules under water sorption from the vapor phase at 20°C in a whole range of relative humidities. On the other hand, water sorption of PA-6 was shown to increase by approximately 30% as the temperature increases from 25 to 100°C [27]. Hence, the assumption that densely packed regions of PA-6 are accessible to water molecules under sorption of OACs from aqueous solutions at 100°C and desorption on treatment the PA-6/OAC samples in boiling water seems to be quite reasonable.

In connection with this, let us discuss the importance of an exact knowledge whether the densely packed regions are accessible for water molecules with respect to the mechanism of distribution of LMs in the polymer between IB and RB forms. Furthermore, the solution of this problem seems to be rather interesting from the viewpoint that according to [27] the accessibility of amorphous regions of semicrystalline polymers (in particular, PA-6) for water molecules is not constant but depends on sorption temperature.

To gain a deeper insight into the mechanism of distribution of LMs in polymer matrix between two forms, let us consider both cases: densely packed amorphous regions are accessible for water molecules under sorption and desorption (i), and densely packed amorphous regions are unaccessible for water molecules under sorption and desorption (ii).

Let us first assume that the densely packed regions of the polymer amorphous phase are accessible for water molecules under sorption and desorption of LMs. The point is that this version agrees well with the above mechanism of distribution of LMs between IB and RB forms. Aqueous solutions of LMs swell the densely packed regions of polymer matrix and the molecules of LMs are able to penetrate these regions. The treatment of the polymer/LM systems in a pure solvent (water) is also associated with swelling of the above regions in water, however, taking into account the above reasoning, in some cases, no dissociation of the polymer/LM complex takes place.

According to the assumption that densely packed regions are unaccessible for water molecules, the amorphous polymer phase under sorption and desorption of LMs may be visualized as heavily swollen loosely packed regions and "dry" densely packed regions. In this case, diffusion of molecules of LMs into the densely packed regions is carried out through a sharp boundary between swollen and "dry" polymer. Actually, we deal with distribution of LMs between loosely packed water-swollen regions and "dry" densely packed domains. In the case of the PAN/Br_2 and PA-6/OAC systems, this explanation seems to be quite plausible due to a high affinity between polymer and LMs [3, 19].

On further treatment the polymer/LM samples in water, desorption of LMs from the densely packed regions proceeds as distribution of LMs between loosely packed water-swollen regions and densely packed dry regions containing this LM. The LM may be completely extracted by solvent from the densely packed regions only in the case when molecules of this agent show sufficient mobility within such structural units. Such situation is likely to be realized for the PA-6/OAC systems only for the compounds I and II, which are able to produce only RB form. The molecules of compound III appear to be immobilized only in the densely packed regions, and no complete extraction of this agent is allowed.

Hence, there is no conflict between the above speculations and the mechanism of distribution of LMs in the polymer matrix between RB and IB forms. In other words, this mechanism provides a plausible explanation of the formation of IB form of LM in polymer matrix independently of the fact whether densely packed regions are accessible for water molecules or not.

So, the development of IB form of LMs in polymer matrix is associated with peculiar features of complexation between molecules of LMs and polymer chains as a result of structural heterogeneity of the amorphous polymer state. In connection with this, the systems, which show the formation of IB forms only in the densely packed regions of the polymer amorphous phase are of most interest, that is, PAN/Br_2 [3] and PA-6/(OAC III) [2, 5, 19] systems. The point is that in this case we can realize a unique possibility of separate modification of different structural levels in polymer amorphous phase. Actually, sorption of LMs from their aqueous solutions by polymers is associated with penetration of molecules of LMs into loosely and densely packed regions of polymer amorphous phase. In other words, both structural levels appear to be modified. The removal of RB fraction of LMs from the loosely packed regions allows one to obtain the polymer samples, in which only one structural level with dense packing of polymer chains is modified.

To understand the peculiar features and possibilities of such separate modification of structural levels of the polymer amorphous phase, let us carry out a comparative analysis of experimental evidence obtained for the polymer systems based on PA-6 and OAC III.

STRUCTURAL FEATURES OF MECHANICAL AND PHYSICAL BEHAVIOR OF POLYMER/(LOW-MOLECULAR-MASS COMPOUND) SYSTEM

The systems based on PA-6 and OAC III are of great interest from the viewpoint of realization of separate modification of different structural levels in the polymer amorphous phase [2, 19]. The treatment of PA-6 films in aqueous solutions of OAC III is associated with modification of both structural levels (Samples I). The removal of RB form from the loosely packed regions implies that only one structural level, that is, densely packed regions, appears to be modified (Samples II).

The temperature dependences of storage modulus E' of dry samples of initial PA-6 samples and Samples I and II (Fig. 1) allows one to outline the following general features of physical and mechanical behavior of such systems [5].

First, the temperature dependences are characterized by the existence of the inversion point of elastic properties of the materials, T_I. At temperatures below T_I, OAC III shows reinforcing action on the polymer elastic properties, whereas at temperatures above T_I this action is weakening.

Second, the removal of RB form from PA-6 (transition from curve 2 to curve 3) is accompanied by shifting T_I towards higher temperatures.

The above behavior was described in [5] in terms of polymer antiplasticization-plasticization with LMs [28, 29]. Molecular mobility of LMs (for example, OAC III) in polymer increases with temperature. An increased molecular mobility of LMs is associated with a decrease in energy of intermolecular interaction between polymer chains, and this provides the conditions for the transition from reinforcing (antiplasticizing) action to weakening (plasticizing) action of OAC III on PA-6. The macroscopic manifestation of this behavior is associated with the appearance of inversion point T_I in the corresponding temperature dependence of elastic properties of polymer systems.

Examination of the behavior of elastic properties of the PA-6/(OAC III) systems in a temperature range from 100 - 110 to 140 - 150°C is of particular interest. In the case of Samples I containing OAC III in both IB and RB forms, the corresponding elastic modulus E' is lower than that of initial polymer sample (curve 2) in this temperature range. For samples II containing only IB form of LMs, the values of E' appear to be higher than those of initial PA-6 samples.

From the viewpoint of molecular dynamics, this experimental evidence suggest that in the above temperature range the molecular mobility of RB form is well-pronounced. Consequently, the RB molecules are able to plasticize the polymer, and a concomitant decrease in elastic modulus is observed. In contrary, molecular mobility of IB form is still frozen-in and its action on polymer matrix is antiplasticizing (reinforcing). In other words, the mobility of RB fraction of OAC III appears at temperatures by 40 - 50°C lower as compared with IB form.

According to our theoretical speculations, the difference between the molecular mobilities of RB and IB forms in a certain temperature range may be rationalized by the fact that IB form is localized in the densely packed domains of the amorphous phase, which are characterized by relatively high level of intermolecular interaction. In connection with this, manifestation of molecular mobility of IB form of OAC is allowed only at elevated temperatures. As the RB fraction of OAC occupy only loosely packed regions of the amorphous phase, which are characterized by lower energetics of intermolecular interaction, then the molecules of this RB fraction acquire mobility at much lower temperatures.

The above evidence allowed us to draw the following conclusions concerning the specific character of elastic behavior of the PA-6/(OAC III) samples. In the aforementioned temperature interval, the situations is realized when one and the same modifying agent is able to have quite different (even opposite) effect on the elastic properties of the material, depending on the form of its existence within the polymer matrix. In other words, in a given temperature range, elastic properties of the PA-6/(OAC III) samples containing OAC in both RB and IB forms are controlled by interplay between plasticizing action of RB form and antiplasticizing action of IB form. The net result depends on the ratio between IB and RB forms of OAC in polymer matrix [5]. This ratio may be changed either by varying the concentration-time regimes of introduction of OAC into the polymer matrix or by desorption of RB fraction of LM from PA-6 [2].

Taking into account the above speculations, one may easily explain the changes in glass transition temperatures T_g of PA-6 samples on sorption and desorption of OAC III (Table 1) [5]. Introduction of OAC into the polymer matrix is accompanied by increase in glass transition temperature T_g of the polymer (Samples I, Table 1). This is main difference between the systems studied and traditional plasticized polymer materials [28, 29]. The removal of RB form of OAC III from the test samples, which is known to exert plasticizing action on the polymer at temperatures above 100°C, i.e., transition from Samples I to Samples II, is accompanied by a growth in T_g of PA-6 samples (Table 1).

Note, that, for oriented PA-6 samples containing OAC III, temperature-induced shrinkage is also controlled by distribution of OAC III between IB and RB forms (Fig. 2). For oriented initial PA-6 films (curves 1), at temperatures below melting point T_m (≈500K), temperature-induced shrinkage, which is characterized by decreasing residual deformation ε_{res}, is likely to be controlled by the relaxation of deformation within amorphous regions of polymer. With increasing temperature close to T_m residual deformation dramatically falls down because of crystallites melting.

At lower temperatures, introduction of OAC III into oriented PA-6 films is responsible for suppression of shrinkage of oriented PA-6 samples containing OAC III (curves 2). As temperature increases the shrinkage of oriented PA-6/(OAC III) samples becomes higher as compared with initial polymer. In other words, for initial oriented PA-6 and oriented PA-6 modified with OAC III, the inversion in shrinkage is observed. As for inversion in elastic properties of isotropic PA-6 films and PA-6/(OAC III) samples, the inversion in shrinkage of oriented polymer samples may be explained by molecular mobility of OAC III in polymer matrix. With increasing temperature molecular mobility of OAC III in PA-6 increases, and decreasing energy of intermolecular interaction between polymer macrochains takes place. As a result, for PA-6/(OAC III) oriented samples, temperature-induced shrinkage is higher as compared with initial PA-6.

For oriented PA-6 samples treated with aqueous OAC III solution with concentration of 0.07 mol/l, at temperatures above 450 K, removal of RB form of OAC III is accompanied with suppression in shrinkage as compared with initial PA-6 and PA-6 containing both IB and RB forms of OAC III (Fig. 2A). At lower temperatures, for oriented PA-6 sample containing only IB form of OAC III (curve 3), temperature-induced shrinkage is higher than that for PA-6 sample containing both IB and RB forms of OAC III (curve 2). Note, that for oriented PA-6 samples treated with aqueous OAC III solution with concentration of 0.007 mol/l, removal of RB form of OAC III results in suppression in shrinkage within the whole temperature range not above melting point (Fig. 2B).

So, examination of physical and mechanical behavior of the PA-6/(OAC III) samples together with speculations concerning the structural heterogeneity of the polymer amorphous phase may serve as a convincing demonstration of possibilities of controlled modification of the bulk properties of polymer materials via controlled modification of different structural levels of amorphous phase of semicrystalline polymer. This conclusion is also supported by comparing the properties of Samples I and II in humid atmosphere (Table 2).

In the case of aliphatic polyamides, sorption of water vapors is known to be accompanied by decrease in elastic modulus of the polymer samples [30, 31]. As compared with initial PA-6, the PA-6/(OAC III) samples containing both IB and RB forms of OAC III (sample I, Table 2) are characterized by a sharp decrease in water sorption and increase in elastic modulus [2]. Desorption of RB fraction of OAC III from PA-6 (transition from Samples I to Samples II, Table 2) is accompanied by an abrupt increase in water sorption, which is comparable to that of initial polymer. However, no decrease in elastic modulus is observed. In other words, distribution of OAC III between IB and RB forms is associated with peculiar "division of labor": RB fraction of modifying agent occupies loosely packed structural levels of polymer amorphous phase and is responsible for water sorption of the polymer material, whereas macroscopic elastic properties of the materials in humid atmosphere are controlled by IB fraction localized in the densely packed regions. Furthermore, such complex modification results in reduced susceptibility of the resultant materials to the plasticizing action of water as evidenced by constancy of elastic modulus even at two-fold increase in water sorption when passing from sample I to sample II (Table 2).

The above evidence on the effect of IB fraction of OAC III localized in the densely packed domains of amorphous phase on macroscopic elastic properties of the polymer samples allows one to draw an important conclusion concerning a marked role of such domains in development of mechanical properties of semicrystalline polymers.

CONCLUSION

Experimental evidence discussed in this work unequivocally shows that in some cases a complicated character of interaction of LMs with polymers is realized. The LMs introduced into the polymer matrix are shown to be distributed between two different forms: reversible (RB) and irreversibly bound (IB) forms. All systems, which show similar behavior, are characterized by high affinity between LMs and polymers, which is controlled by complexation between molecules of LM and polymer chains.

The mechanism of distribution of LMs between IB and RB forms is based on speculations concerning structural inhomogeneity of amorphous component in semicrystalline or completely amorphous polymers: the existence of noncrystalline regions (structural levels) with different packing density and order. The development of IB form of LMs is a result of complexation reaction between molecules of LM and polymer chains and formation of very stable complex. The possibility of occurrence of such complexation reactions is dictated by chemical structure of LM and local packing density of polymer chains. In connection with this, IB form may be localized within different structural levels or even does not form at all.

In turn, the complicated character of interaction between LMs and polymer chains is responsible for a complicated molecular dynamics of polymer/LM systems and, hence, for specific features of mechanical behavior of the material.

The approaches discussed in this work may be interpreted as underlying principles for new scientific direction, according to which interaction between LMs and polymers is primarily treated from the viewpoint of existence of different levels of structural inhomogeneity in polymers. This direction involves some general problems related to sorption and diffusion of LMs in polymer bodies with a complicated structural organization.

On the other hand, controlled occupation of different levels of structural inhomogeneity with molecules of LM suggests many possibilities of controlled modification and improvement in mechanical properties of polymer materials.

Table 1

Glass Transition Temperatures of Dry PA-6/(OAC III) Samples

Sample	Content of OAC III in PA-6, mol/kg	T_g,K
Initial PA-6	0	353
Sample I	0.29	388
Sample II	0.09	396
Sample I	0.50	393
Sample II	0.16	406

Table 2

Elastic Properties of the PA-6/(OAC III) Samples in Humid Atmosphere

Sample	Content of OAC III in PA-6, mol/kg	p/p_o,%	Water content, mol/kg	Elastic Modulus, GPa
Initial PA-6	0	0	0	1.9
	0	47	2.8	1.3
Sample I	0.50	47	0.95	2.5
Sample II	0.16	47	2.4	2.5

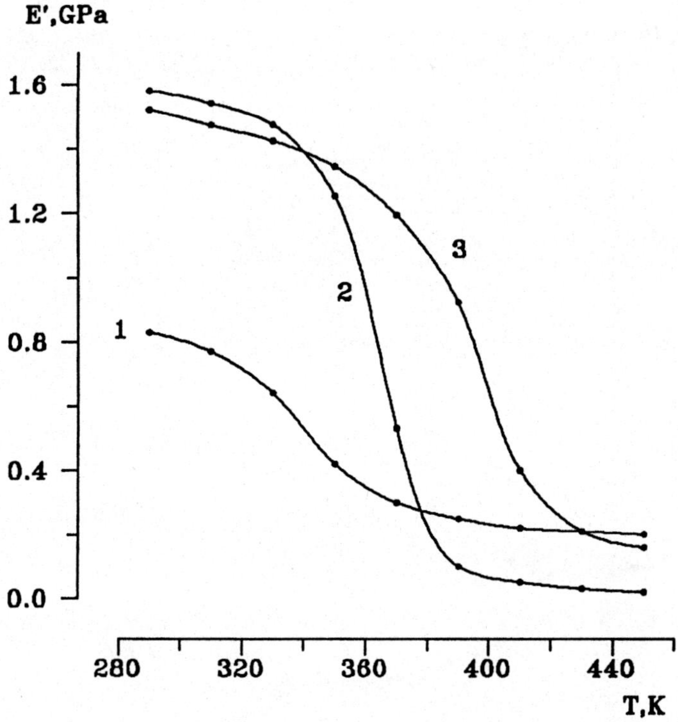

Fig. 1. Temperature dependences of storage moduli E' of PA-6 (1) and PA-6/(OAC III) samples. OAC content: 0.5 mol/kg (Sample I) (2), and 0.16 mol/kg (Sample II) (3).

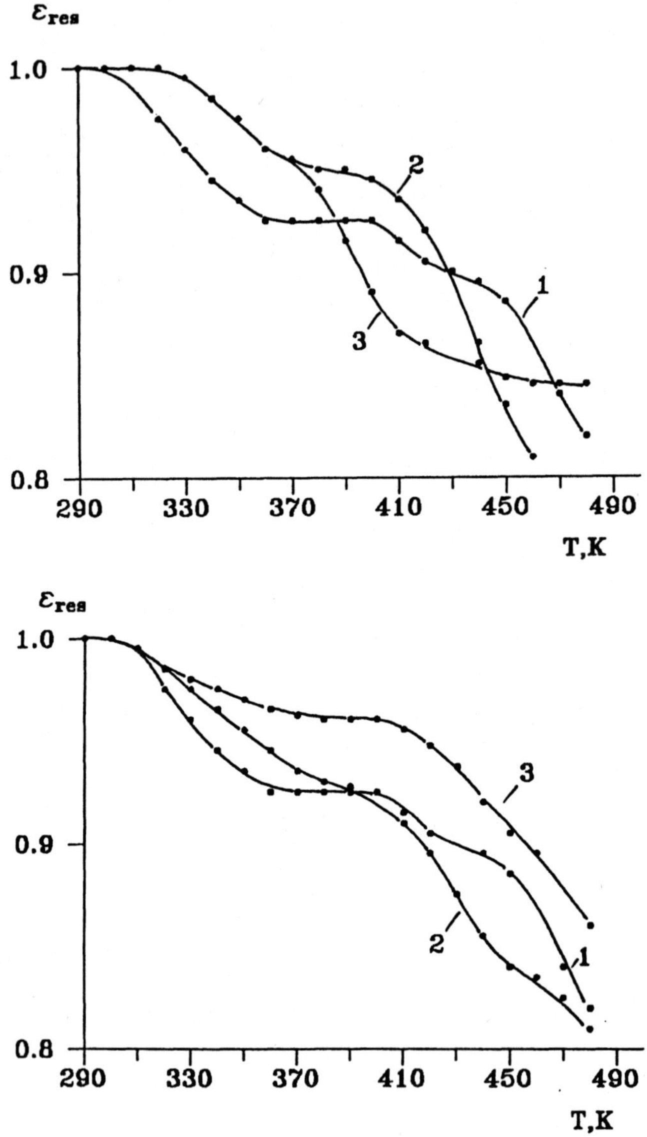

Fig. 2. Temperature dependences of relative residual deformation ε_{res} of oriented PA-6 film (1), oriented PA-6 treated with aqueous OAC solutions with concentrations 0.07 (A) and 0.007 mol/l (B) (2), and, then, with boiling water for 6 h (3).

REFERENCES

1. M.S. Arzhakov, A.V. Volkov, A.L. Volynskii and N.F. Bakeev, Makromol. Chem., Macromol. Symp., 44, 165 (1991).

2. M.S. Arzhakov, N.V. Red'ko, A.V. Volkov, A.L. Volynskii and N.F. Bakeev, Polym. Sci., A, 36, 1125 (1994).

3. M. Lewin, H. Guttman and Y. Naor, J. Macromol. Sci., A, 1367 (1988).

4. A.Ya. Polishchuk, L.A. Zimina, N.N. Madyuskin and G.E. Zaikov, Polym. Sci., A, 35, 80 (1993).

5. M.S. Arzhakov, E.V. Nechaeva, A.V. Volkov, A.L. Volynskii and N.F. Bakeev, Polym. Sci., A, 35, 521 (1993).

6. A.V. Volkov, M.S. Arzhakov, V.V. Bondarev, A.L. Volynskii and N.F. Bakeev, Vysokomol. Soedin., A, 32, 467 (1990) (in Russian).

7. A.V. Volkov, M.S. Arzhakov, A.L. Volynskii and N.F. Bakeev, Dokl. Akad. Nauk SSSR, 306, 1401 (1989) (in Russian).

8. J. Guthrie and C.C. Cook, Polymer, 22, 1439 (1981).

9. Y. Kishimoto, T. Hosaka and W. Shimotsuma, J. Appl. Polym. Sci., 32, 5161 (1986).

10. M. Topic, A. Mogus-Milankovic and Z. Katovic, Polymer, 28, 33 (1987).

11. A.L. Volynskii, A.V. Volkov, V.V. Bondarev, N.I. Seleznev, M.S. Arzhakov and N.F. Bakeev, Vysokomol. Soedin., A, 32, 2323 (1990) (in Russian).

12. D.C. Prevorsec, J. Polym. Sci., C, 32, 343 (1971).

13. D.C. Prevorsec, G.A. Tirpac, P.J. Harget and H.C. Reimschuessel, J. Macromol. Sci., B, 9, 733 (1974).

14. G.S.J. Yeh, J. Macromol. Sci., B, 6, 465 (1972).

15. I.J. Klement and P.H. Geil, J. Macromol. Sci., B, 5, 505 (1971).

16. S.A. Arzhakov, N.F. Bakeev and V.A. Kabanov, Vysokomol. Soedin, A, 15, 1154 (1973) (in Russian).

17. G.V. Kozlov, D.S. Sanditov and V.D. Serdyuk, Polym. Sci., A, 35, 1742 (1993).

18. I.I. Perepechko and S.S. Yakovenko, Vysokomol. Soedin., A, 23, 1166, (1981) (in Russian).

19. M.S. Arzhakov, E.V. Nechaeva, N.V. Red'ko, A.V. Volkov, A.L. Volynskii and N.F. Bakeev, Polym. Sci., A, 38, 62 (1996).

20. A.V. Volkov, M.S. Arzhakov, A.L. Volynskii and N.F. Bakeev, Vysokomol. Soedin., A, 32, 489 (1990) (in Russian).

21. R.M. Myasnikova, V.D. Samarskaya and E.S. Obolonkova, Vysokomol. Soedin., B, 34, 43 (1992) (in Russian).

22. R.M. Myasnikova, E.F. Titova and E.S. Obolonkova, Polymer, 21, 403 (1980).

23. T.P. Yang, E.M. Pearce, T.K. Kwei and N.L. Yang, Macromolecules, 22, 1813 (1989).

24. V.A. Kabanov and I.M. Papisov, Vysokomol. Soedin., A, 21, 243 (1979) (in Russian).

25. I Iliopoulos and R. Audebert, J. Polym. Sci., B, 26, 2093 (1988).

26. M.S. Arzhakov, Cand. Sci. (Chem.) Dissertation, Moscow: Moscow State University, 1989 (in Russian).

27. L.P. Razumovskii, V.S. Markin and G.E. Zaikov, Vysokomol. Soedin., A, 27, 675 (1985) (in Russian).

28. P.V. Kozlov and S.P. Papkov, Fiziko-Khimicheskie Osnovy Plastifikatsii Polimerov (Physicochemical Foundations of Plasticization of Polymers), Moscow: Khimiya, 1982 (in Russian).

29. L.E. Nielsen, Mechanical Properties of Polymers and Composites, New York: Marcel Dekker, 1974.

30. Water in Polymers, Ed., S.P. Rowland, Washington, D.C., 1980.

31. R. Puffr and J. Sebenda, J. Polym. Sci., C, 79 (1967).

Interaction of Polyamide-6 with Mixtures of Oxyaromatic Compounds

M.S. Arzhakov, E.V. Nechaeva, N.V. Red'ko, A.V. Volkov, L.A. Kazarin, A.L. Volynskii and N.F. Bakeev

Polymer Department, Faculty of Chemistry, Lomonosov Moscow State University, Vorob'evy Gory, Moscow, 119899 Russia

Interaction of polyamide-6 with the mixtures of oxyaromatic compounds introduced in the amorphous regions of polymer from aqueous solutions is shown to be controlled by the competing complexation reactions between the mixture components with polymer macromolecules. The principal contribution to binding with polymer matrix is provided by the component, which is able to produce a cooperative set of intermolecular hydrogen bonds with polyamide-6 macrochains. The development of the cooperative intermolecular bonds between the molecules of oxyaromatic compounds and polymer macromolecules is discussed in terms of structural inhomogeneity of amorphous phase of polyamide-6: co-existence of structural sublevels with different packing densities.

Key words: PA-6, oxyaromatic compounds, interaction, structure

INTRODUCTION

In previous papers [1 - 3] we showed that, in some cases, introduction of oxyaromatic compounds (OACs) to PA-6 from their aqueous solutions is accompanied by the distribution of these low-molecular-mass agents in polymer matrix between reversibly bound (RB) and irreversibly bound (IB) forms. As compared with IB form , the RB form of OAC may be easily removed from polymer by treatment of PA-6 with OAC in pure solvent (water). The similar distribution of low-molecular-mass compounds between IB and RB forms was observed for Br_2 in poly(acrylonitrile) [4] and dioxydine in copolymer of N-vinylpyrrolidone with methyl methacrylate [5].

In [6, 7], the development of IB form of OAC in PA-6 was discussed in terms of complexation between molecules of OACs and polymer macromolecules taking into consideration structural inhomogeneity of amorphous regions of PA-6, that is, co-existence of noncrystalline structural levels with different packing densities and ordering. Let us note that structural inhomogeneity of amorphous phase of semicrystalline polymers, in particular, polyamides, was widely discussed in literature (for example, in [8 - 10]).

The formation of IB form of OAC was shown to be controlled by the development of a cooperative set of hydrogen bonds between phenol hydroxyls of OACs and amide groups of polymer. Depending on chemical structure of OAC molecules, IB portion of OAC may be produced both in the loosely packed structural sublevels and in the densely packed structural sublevels or may not be formed.

The aim of the present work was to investigate the distribution of OACs between IB and RB forms during the introduction of mixtures of these low-molecular-mass agents into PA-6.

EXPERIMENTAL

We studied the commercial isotropic films of PA-6 (trademark PK-4, Russia) with a thickness of 0.1 mm. Before testing, the samples were treated in boiling water for 3 h. After that, they were dried over P_2O_5 until constant mass was attained.

As OACs we used dioxydiphenylsulfone (I), synthetic tanning agent (II), which was prepared by condensation of sulfonated β-naphthol and dioxydiphenylsulfone with formaldehyde, and resorcinol-formaldehyde novolac (III). The low-molecular-mass compounds were purified using dialysis. The purity of OACs was controlled by IR-spectroscopy.

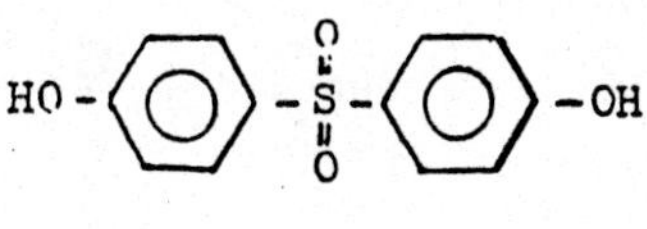

I

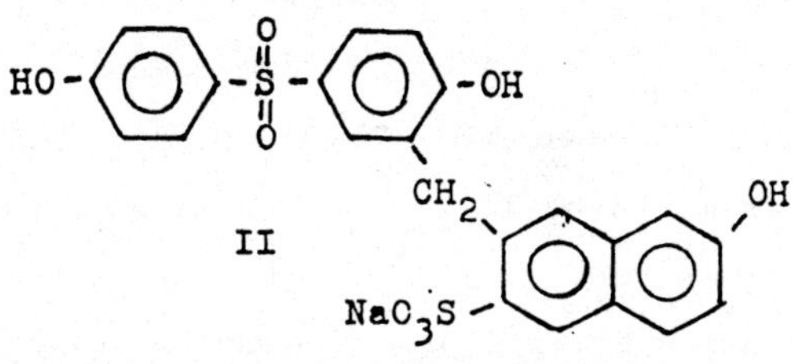

II

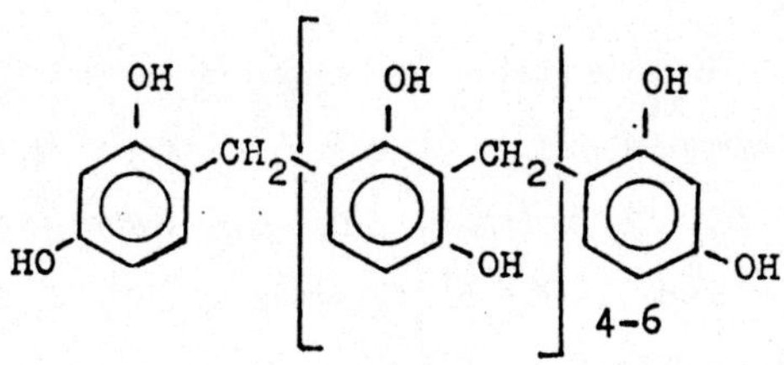

III

The films of PA-6 were treated at 100°C in aqueous solutions of OACs with concentrations: 0.0125 and 0.025 (OAC I), 0.0125 and 0.025 (OAC II), 0.025 (OAC III) mol/l, as well as in aqueous solutions of mixtures of OAC II with OAC I and OAC III with OAC I until equilibrium sorption of OACs by polymer was attained. Molar fraction of OAC III in its mixture with OAC I was 0.5 (mixture of 0.025 mol of OAC III with 0.025 mol OAC I was dissolved in 1 l of water). Molar fractions of OAC II in its mixture with OAC I were 0.33 (mixture of 0.0125 mol OAC I with 0.025 mol OAC I was dissolved in 1 l of water), 0.5 (mixture of 0.025 mol OAC I with 0.025 mol OAC I was dissolved in 1 l of water), and 0.66 (mixture of 0.025 mol OAC I with 0.0125 mol OAC I was dissolved in 1 l of water). After the treatment of PA-6 films in aqueous solutions of OACs, the samples were dried over P_2O_5 until constant mass was attained. Concentration of OACs (or their mixtures) sorbed by the polymer samples was estimated gravimetrically.

To obtain quantitative estimates of OACs introduced into PA-6 films between RB and IB forms, the PA-6 films with OACs were treated in boiling water until equilibrium desorption of OACs from polymer was attained. The IB form of OAC was defined as the fraction of OAC that cannot be removed from PA-6 samples by their treatment in boiling water.

The estimation of the composition of mixture of OACs, which may be desorbed from PA-6 samples after their treatment in aqueous solutions of mixtures of OAC I and OAC II, was carried out as follows. The water, in which PA-6 films containing OACs were treated, was evaporated using water bath. After that, the pellet of dry residue with KBr was prepared. IR-spectra of this pellet were taken using a Specord M-80 instrument (Germany). The content of OAC II was estimated from the ratios of intensities of 840, 900, and 1030 cm^{-1} bands according to [11].

RESULTS AND DISCUSSION

Figure 1 shows the kinetic curves describing desorption of OACs from PA-6 films. This experimental evidence suggests that, for individual OACs I - III, the formation of IB portion is observed for OACs II (curves 3 and 4) and III (curve 5). The fractional contents of IB form in the total content of these OACs in polymer are 0.32±0.02 and 0.78±0.02, respectively. OAC I is not able to produce in PA-6 IB form and may be completely desorbed from polymer (curves 1 and 2).

Desorption of OACs from PA-6 films treated in aqueous solutions of mixture of OAC II with OAC I and mixture of OAC III with OAC I (curves 6 - 9) evidences that, in all cases, the formation of IB form is observed. The above results on desorption of individual OACs from polymer allow one to conclude that only OACs II and III are able to produce IB form during the sorption of their mixtures with OAC I.

After the treatment of PA-6 films in aqueous solution of mixture of OAC III with OAC I, the fractional content of IB form in polymer matrix is equal to that as observed after the treatment of PA-6 films with aqueous solution of OAC III (curves 5 and 9). Hence, the presence of OAC I in mixture has no effect on the distribution of OAC III between IB and RB forms in the amorphous regions of PA-6.

In the case of PA-6 films treated with aqueous solutions of

mixtures of OAC II with OAC I, as the concentration of OAC I in mixture increases, fractional content of IB form of OAC II decreases (curves 6 - 8). To obtain a detailed information concerning the effect of OAC I on the distribution of OAC II in PA-6 between the IB and RB forms, the fractional content of RB form of OAC II in polymer matrix was estimated.

To this end, the composition of mixture of OAC II and OAC I desorbed from PA-6 was examined using IR-spectroscopy. Figure 2 shows the corresponding IR-spectra of OAC I and OAC II (curves 1 and 2, respectively), as well as the IR-spectra of low-molecular-mass compounds desorbed from PA-6 films treated with aqueous solutions of mixtures of OAC II with OAC I (curves 3 - 5). OAC II may be identified from the 900 and 1030 cm^{-1} bands. These absorption IR bands are associated with $R\text{-}SO_2OH$ groups or related salts [12]. With increasing the concentration of OAC II in aqueous solutions of OAC II with OAC I, the intensities of 900 and 1030 cm^{-1} bands increase. This experimental evidence allows one to conclude that the concentration of OAC II in the mixtures of OAC II with OAC I desorbed from PA-6 increases. The concentration of OAC II in mixture of OAC II with OAC I desorbed from PA-6, that is, the content of RB form of OAC II in polymer, may be estimated from the ratio between the intensities of 900 and 1030 cm^{-1} bands and intensity of 840 cm^{-1} band. Let us note that the latter band is characteristic of both OAC II and OAC I. Molar fractions of OAC II as well as its IB and RB forms are presented in Table 1.

This evidence suggests that, at first, the dependence of molar fraction of OAC II in its mixture with OAC I in polymer on molar fraction of OAC II in its mixture with OAC I in aqueous solutions does not show additivity (Fig. 3, curve 1). This implies that the preferential sorption of OAC I from the aqueous solutions of mixtures of OAC I with OAC II by PA-6 takes place. Secondly, OAC II introduced to PA-6 from aqueous solutions of mixtures of OAC II with OAC I primarily exists in polymer in its IB form. The fractional content of IB OAC II, with respect to the total content of OAC II in polymer, [IB]/([RB]+[IB]) is 0.66±0.05, independently of the concentration of OAC II in aqueous solutions of its mixture with OAC I. Let us note that, for sorption of individual OAC II by polymer, the corresponding value is 0.32±0.02. Deviation of dependence of molar fraction of IB OAC II in mixture of OAC II with OAC I in polymer on molar fraction of OAC II in its mixture with OAC I in solution from additivity is not noticeable (Fig. 3, curve 2).

The development of IB form of OAC in PA-6 is controlled by the specific features of complexation of OAC molecules with polyamide macromolecules in the amorphous phase of polymer [6, 7]. These features are associated with structural inhomogeneity of amorphous phase, that is, co-existence of, at least, two different noncrystalline levels with different packing densities. The development of IB OAC is controlled by the complexation between OAC and PA-6, and the as-formed complex is stable to action of solvent (in this case, water). The development of stable complexes between low-molecular-mass compounds and macromolecules is observed when cooperativity of a set of intermolecular bonds takes place, and joint dense packing of the molecules of the reacting compounds is realized. The fulfillment of these conditions in the amorphous phase of PA-6 is decided by the chemical structure of molecules of low-molecular-mass compounds (molecular size and number of complexing groups) and local packing of macromolecules.

In the case of OAC II, cooperative interaction with PA-6 macromolecules, that is, the development of IB form, takes place only in the densely packed regions of the amorphous phase of PA-6. In the loosely packed regions, OAC II exists in its RB form and may be easily removed from polymer. When passing to OAC III, with increasing the molecular size and number of complexing hydroxyl groups, the development of IB form also in the loosely packed regions of the amorphous phase of PA-6 takes place. For OAC I, the development of only RB form suggests that no cooperative interaction of the molecules of this low-molecular-mass compound with PA-6 macromolecules is observed in the whole amorphous phase.

When the mixtures of OACs are sorbed by PA-6, both in loosely and densely packed structural regions there is a competition between the components of mixture in complexation with PA-6 macromolecules. In this case, the component of mixture that is able to produce a cooperative set of intermolecular bonds with PA-6 macromolecules has more advantages in binding with polymer [13, 14]. In the case of the mixture of OAC I with OAC III, only the latter is capable producing cooperative interaction with polymer macrochains in amorphous regions of PA-6 [7]. In connection with this, preferential binding of OAC III with PA-6 is observed. Let us note that OAC I seems to have no effect on the complexation between OAC III and PA-6 macromolecules as well as on the distribution of OAC III between IB and RB forms (Fig. 1, curves 5 and 9).

In the case of mixtures of OAC I with OAC II, the latter component is able to produce a cooperative interaction with PA-6 macromolecules but only in the densely packed structural sublevels of the amorphous phase of polymer [7]. Within these structural sublevels, as evidenced by slight deviation of curve 2 (Fig. 3) from additivity the preferential binding of OAC II by PA-6 is observed.

Within the loosely packed structural regions in the amorphous phase of PA-6, for both OAC I and OAC II no development of cooperative interaction with polymer macromolecules is observed. Within these structural sublevels, these low-molecular-mass compounds exist in their RB form. Nevertheless, as evidenced by rather low molar fraction of RB OAC II (Table 1) the loosely packed regions are enriched with OAC I. This evidence may be rationalized by the fact that molecules of OAC II involve ionogenic SO_3Na groups, which as compared with OAC I provide a better affinity of this low-moleoular-mass compound toward water. As a result, the sorption of low-molecular-mass compounds from aqueous solutions of mixtures of OAC II with OAC I is accompanied by the enrichment of the loosely packed structural regions with OAC I.

Hence, the sorption of mixtures of OAC I with OAC II by PA-6 is accompanied by the distribution of the components of mixtures between different structural regions in the amorphous phase of polymer. Within the loosely packed regions, the preferential binding of OAC I is observed. On contrary, OAC II is primarily sorbed by the densely packed regions. Eventually, as evidenced by the deviation of curve 1 (Fig. 3) from additivity amorphous phase of PA-6 is enriched by OAC I.

CONCLUSION

Sorption of mixtures of OACs from their aqueous solutions is associated with the competing in complexation reactions between the mixture components with polymer macromolecules. The component of mixture that is able to produce a cooperative set of intermolecular bonds with PA-6 macrochains is predominantly bound by polymer. The development of the cooperative set of intermolecular bonds is controlled by the chemical structure of molecules of OAC (molecular size and number of complexing groups) and local packing of macromolecules. These factors are responsible for the enrichment of structural regions of amorphous phase of PA-6 with different packing densities with the certain component of mixture.

Table 1

Concentration of OAC II and fractional content of its IB and RB forms in mixture with OAC I in PA-6 as a function of concentration of OAC II in mixture with OAC I in aqueous solutions

Molar fraction of OAC II in mixture with OAC I in aqueous solution	Molar fraction of OAC II in mixture with OAC I in polymer matrix			$\frac{[IB]}{[RB]+[IB]}$
	total	IB form	RB form	
1.0	1.0	0.32	0.68	0.32
0.66	0.23	0.14	0.09	0.61
0.5	0.17	0.12	0.05	0.71
0.33	0.12	0.08	0.04	0.67
0	0	0	0	

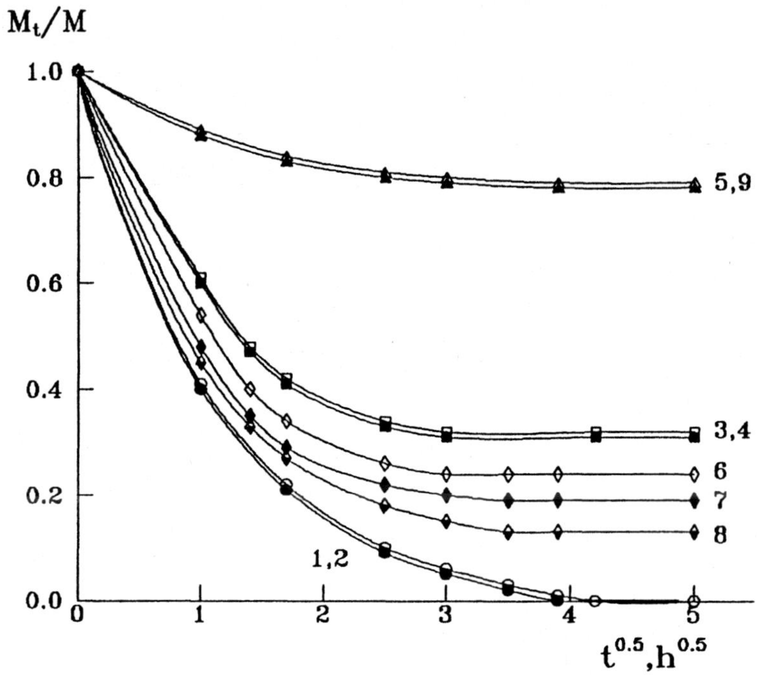

Fig. 1. Kinetic curves of desorption of OACs from PA-6.

M_t - content of OAC in PA-6 after desorption for a time t, M - equilibrium content of OACs in PA-6 treated with aqueous solutions of OAC I with concentrations 0.0125 (1) and 0.025 (2), OAC II with concentrations 0.0125 (3) and 0.025 (4), OAC III with concentration 0.025 (5) mol/l, mixtures of OAC II with OAC I (molar fraction of OAC II was 0.66 (6), 0.5 (7), and 0.33 (8)), and mixture of OAC III with OAC I (molar fraction of OAC III was 0.5 (9)).

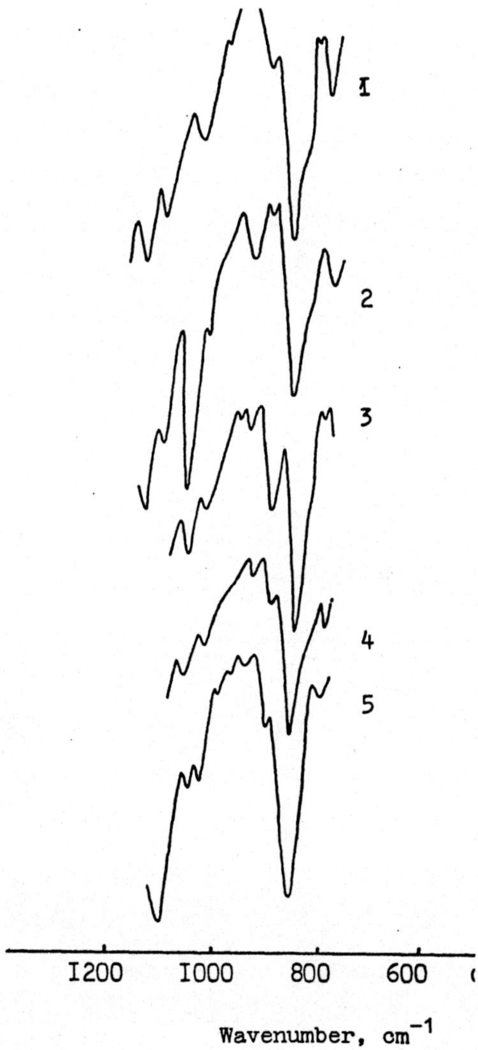

Fig.2. The absorption IR-spectra of OAC I (1) and OAC II (2) as well as of residues desorbed from PA-6 films after their treatment with aqueous solutions of mixtures of OAC II with OAC I. Molar fractions of OAC II in mixture with OAC I were 0.66 (3), 0.5 (4), and 0.33 (5).

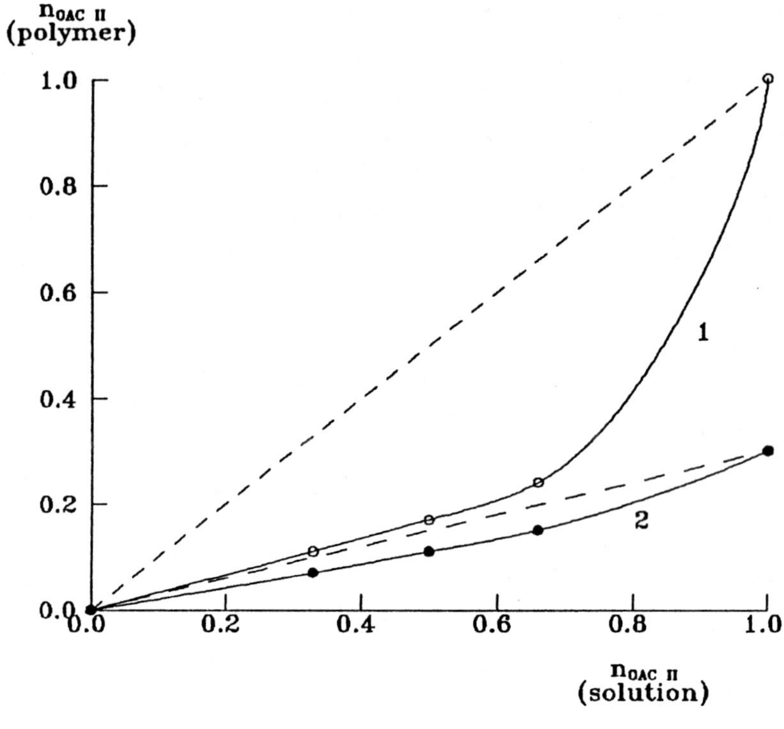

Fig. 3. Molar fraction of OAC II in mixture with OAC I in PA-6 (1) and molar fraction of IB OAC II in mixture of OAC II with OAC I in PA-6 (2) versus molar fraction of OAC II in mixture with OAC I in aqueous solution. Dashed lines correspond to additivity.

REFERENCES

1. M.S. Arzhakov, A.V. Volkov, A.L. Volynskii and N.F. Bakeev, Makromol. Chem., Macromol. Symp., 44, 165 (1991).

2. M.S. Arzhakov, E.V. Nechaeva, A.V. Volkov, A.L. Volynskii and N.F. Bakeev, Polymer Science, A, 35, 521 (1993).

3. M.S. Arzhakov, N.V. Red'ko, A.V. Volkov, A.L. Volynskii and N.F. Bakeev, Polymer Science, 36, 929 (1994).

4. M. Lewin, H. Guttman and Y. Naor, J. Macromol. Sci., Chem., 1367 (1988).

5. A.Ya. Polishchuk, L.A. Zimina, N.N. Madyuskin and G.E. Zaikov, Polymer Science, A, 35, 80 (1993).

6. M.S. Arzhakov, E.V. Nechaeva, N.V. Red'ko, A.V. Volkov, A.L. Volynskii and N.F. Bakeev, Polymer Science, A, 37, 71 (1996).

7. M.S. Arzhakov, Intern. J. Polymeric Mater., 28, 103 (1995).

8. I.I. Perepechko and S.S. Yakovenko, Vysokomol. Soedin., A, 23, 1166 (1981) (in Russian).

9. N.S. Murphy and W.J. Orts, J. Polym. Sci., Polym. Phys., 32, 2695 (1994).

10. N.S. Murphy, M. Stamm, J.P. Sibilia and S. Krimm, Macromolecules, 22, 1261 (1989).

11. J. Dechant,R. Danz, W. Kimmer and R. Schmoltke, Ultrarotspektroskopische Untersuchungen an Polymeren, Berlin: Akademie Verlag, 1972.

12. K. Nakanishi, Infrared Absorption Spectroscopy, San-Francisco: Holden-Day, Inc., Tokyo: Nankado Company, Ltd., 1962.

13. V.A. Kabanov and I.M. Papisov, Vysokomol. Soedin., A, 21, 243 (1979) (in Russian).

14. T.P. Yang, E.M. Pearce, T.K. Kwei and N.L. Yang, Macromolecules, 22, 1813 (1989).

Physical and Mechanical Behavior of Polyamide-6 containing Oxyaromatic Compounds

M.S. Arzhakov, E.V. Nechaeva, I.I. Khairullin,
A.V. Volkov, A.L. Volynskii and N.F. Bakeev

Polymer Department, Faculty of Chemistry, Moscow State University, Vorob'evy Gory, Moscow, 119899 Russia

The distribution of oxyaromatic compound in PA-6 amorphous regions between two structural forms, which are characterized by different energy of interaction with polymer matrix, was studied. Physical and mechanical properties (glass transition temperature, water sorption, elastic modulus) of PA-6 containing oxyaromatic compound were shown to be controlled not only by the net content of oxyaromatic compound in polymer, but by its distribution between these structural forms. This distribution appears to govern the mechanical properties of the polymer in the region of its relaxation transition from glassy to rubbery state, whatever this state was achieved by increasing the concentration of plasticizer (water) or by increasing the temperature. The experimental evidence obtained was discussed with the account for structural heterogeneity of the amorphous regions of semicrystalline PA-6.

Key Words: polyamide-6, oxyaromatic compound, modification, mechanical properties

INTRODUCTION

Improvement of physical and mechanical properties of polymer materials may be achieved by their physical modification via introduction of compatible low-molecular-mass compounds (LMs) into polymers. Generally, for these polymer/LM systems, compatibility of the components is controlled by complexation between polymer macrochains and molecules of LM via hydrogen bonding, electrostatic, dispersion, and similar interactions. Physical and mechanical behavior of as-modified polymer materials is decided by energetics of interaction between components of polymer composite.

In some cases, LMs, which were introduced into the polymer from the swelling solutions, appear to be distributed within polymer matrix between two different forms, reversibly (RB) and irreversibly bound (IB) with polymer [1 - 4]. As compared with IB form, RB form may be easily removed from polymer by keeping polymer/LM samples in a pure solvent. In other words, the above forms of LM are characterized by the different energy of interaction with polymer. As a result, physical and mechanical behavior of these polymer materials are expected to be primarily controlled by the distribution of modifying agent between RB and IB forms.

Mechanism of the distribution observed was discussed [1, 2, 5, 6] in terms of structural inhomogeneity of amorphous polymers or amorphous phase in semicrystalline polymers - existence of, at least, two different noncrystalline regions with different packing densities and ordering. Note, that the structural inhomogeneity of the amorphous phase in semicrystalline polymers and completely amorphous polymers is an intriguing problem, which is being vigorously discussed in literature [7 - 15]. According to mechanism proposed, development of IB form of LM in polymer is associated with the formation of a stable complex between LM molecules and polymer macrochains in densely packed structural regions of polymer. Complexation of LM molecules with polymer macrochains in loosely packed structural regions is accompanied by development of RB form of LM with lowered energy of interaction as compared with IB form. As a result, RB form of LM is able to be removed from polymer matrix. This approach allows controlled occupation of different structural regions of polymer with LM molecules, i.e., separate modification of the above structural regions.

In this paper, for model systems based on PA-6 modified with oxyaromatic compound (OAC), the effect of distribution of modifying agent between loosely and densely packed structural regions within amorphous phase of polymer on physical and mechanical behavior of as-modified polymer material was studied. OACs are known [3, 16 - 18] to be well compatible with aliphatic polyamides because of complexation between phenol hydroxyls of OAC with amide polymer groups via hydrogen bonding. This complexation was shown to be responsible for increase in the energy of intermolecular interaction in polyamides, and, as a result, for increasing mechanical characteristics of polymer [19, 20]. Distribution of OACs within amorphous phase of aliphatic polyamides was noted in [4 - 6]. Moreover, the ratio between IB and RB forms of OAC in PA-6 was shown to be controlled by changing concentration and time regimes of the treatment of PA-6 samples in swelling OAC solutions.

EXPERIMENTAL

We studied commercial isotropic films of PA-6 with a thickness of 0.1 mm. Before testing, the samples were treated in boiling water for 3 h. After that, they were dried over P_2O_5 until constant mass was attained. As a modifying agent, we used an oxyaromatic compound prepared by condensation of sulfonated β-naphtol and dioxydiphenylsulfone with formaldehyde, with the following structural formula:

$$HOSO_2(C_{10}H_5OH)CH_2(C_6H_3OH)SO_2(C_6H_4OH)$$

To introduce OAC to PA-6, polymer films were treated with OAC aqueous solutions with concentrations of 0.007, 0.036, 0.072, and 0.14 mol/l for 5, 15, 30, 60, and 180 min at 373 K. Then, the PA-6/OAC samples were allowed to dry over P_2O_5 until constant mass was attained. Concentration of OAC sorbed by the polymer samples was estimated gravimetrically.

The kinetic sorption curves (Fig. 1) show that equilibrium sorption of OAC by PA-6 samples is achieved within 3 h. With increasing the time of treatment to 6 h, no further sorption of OAC by PA-6 films was observed.

To obtain quantitative estimations on distribution of OAC introduced to PA-6 films between RB and IB forms, the PA-6/OAC samples were treated in boiling water for 1, 2, 3, 6, and 12 h. We showed that RB form of OAC can be removed from the polymer within 6 h. With increasing the time of treatment to 12 h, no marked desorption of OAC from PA-6 was observed. In this connection, the IB form of OAC can be defined as the fraction of OAC, which cannot be removed from the PA-6/OAC samples upon their treatment in boiling water for 6 h. The fractional content of IB form of OAC was estimated as C_{IR}=[IR]/([IR]+[RB]).

Figure 2 shows the kinetic curves of accumulation of IB form of OAC in amorphous regions of PA-6. For all aqueous OAC solutions, with increasing the time of treatment of PA-6 films to 3 h, fractional content of IB form of OAC approaches the same value (0.32 ± 0.02). This value is likely to be related to equilibrium distribution of OAC between IB and RB forms, since it remains the same even for the PA-6 samples treated with aqueous OAC solutions for 6 h.

Hence, the variations in the net content of OAC in PA-6 as well as in the fractional content of IB form of OAC in polymer samples can be carried out using the following procedures:

1. Treatment of PA-6 films with aqueous OAC solutions of different concentrations for 3 h allows one to vary the net content of OAC at fixed fractional content of its IB form - 0.32 ± 0.02 (Samples I).
2. Treatment of PA-6 films with aqueous OAC solutions of fixed concentration for various periods of time allows one to vary fractional content of IB form of OAC in the range not exceeding 0.32 (Samples II).
3. Removal of RB form of OAC from PA-6/OAC samples allows one to increase fractional content of OAC in polymer to 1.0 (Samples III).

The dynamic properties of PA-6/OAC dry samples were studied with Rheometrics solid analyzer RSA-II (USA) over temperature range from 290 to 450 K; deformation frequency was 5 Hz; strain was less than 0.5%; heating rate was 2 K/min. Glass transition temperatures of PA-6/OAC samples were estimated from temperature dependences of loss tangent.

To study water sorption, the PA-6/OAC samples were conditioned over saturated aqueous solutions of inorganic salts at a given humidity at 293 K until constant mass. Water sorption was measured gravimetrically.

Mechanical tests were carried out with an Instron-1100 tensile machine (Great Britain) at a strain rate of 100 %/min at 293 K. The initial elastic modulus was measured at strains below 3%. Prior to mechanical tests, the samples were conditioned over saturated aqueous solutions of inorganic salts at a given humidity until equilibrium.

Uniaxial drawing of isotropic PA-6 films was carried out in water at 293 K to strain of 160%. After that, the stressed samples were dried in vacuum. Dry stressed PA-6 samples were treated in aqueous OAC solutions with concentrations of 0.007, 0.036, and 0.072 mol/l for 3 h at 373 K. Removal of RB form of OAC was carried out by treatment of stressed oriented PA-6/OAC samples in boiling water for 6 h. Then, oriented PA-6/OAC samples were dried in vacuum and unloaded. Temperature-induced shrinkage of oriented PA-6/OAC samples was studied in the temperature range from 293 K to 483 K. Residual deformation at a given temperature was estimated as $\varepsilon_{res} = \frac{l_T - l_0}{l_{def} - l_0}$, where l_0 - initial length of PA-6 sample, l_{def} - the length of the oriented sample, l_T - the length of the sample at a given temperature.

RESULTS AND DISCUSSION

Figure 3 shows the temperature dependences of storage moduli E' for Samples I. As is visible, the temperature dependences of E' have an inversion point, T_I. That is, at temperatures below T_I, increasing the content of OAC in PA-6/OAC samples leads to an increase in E', whereas at temperatures above T_I with increasing the content of OAC in polymer E' decreases.

Such behavior may be attributed to the antiplasticizing (reinforcing) action of OAC on the mechanical properties of PA-6 at temperatures below T_I and to the plasticizing (weakening) action at temperatures above T_I. Note, that, for Samples I, T_I is independent of the content of OAC in PA-6. A plausible explanation of this phenomenon will be given below.

When RB form of OAC is removed from PA-6/OAC samples by their treatment in boiling water (transition from Samples I to Samples III), T_I is shifted towards higher temperatures (Fig. 4). It is of particular interest to analyze the behavior of the elastic properties of PA-6/OAC samples in the temperature range from 373-383 to 413-423 K.

In this temperature range, the values of storage moduli E' for Sample I (Fig. 4, curve 2), which contains both IB and RB forms of OAC, are lower than those for the initial polymer (curve 1). In the case of Sample III (curve 3), which contains only IB form of OAC, the values of E' are still higher than those for the initial polymer. This means, that in the samples studied, the same modifying agent may have opposite effect on the elastic properties of the polymer, depending on its form in the polymer matrix. Let us discuss the behavior of elastic properties of PA-6/OAC samples in terms of the molecular mobility of OAC in PA-6.

With increasing temperature, the mobility of OAC molecules in PA-6 increases. Increasing mobility of OAC molecules is responsible for decreasing the energy of intermolecular interaction between polymer macromolecules. As a result, antiplasticizing action of OAC on PA-6 is replaced by plasticizing action. For the PA-6/OAC samples, this mechanism may be invoked to explain the existence of the inversion point T_I in the temperature dependences of storage moduli.

The peculiarities of the mechanical behavior of the PA-6/OAC samples are related to their complicated molecular organization, i.e., to the existence of two different forms (IB and RB) of OAC in polymer matrix. In a certain temperature range, mobility of RB form of OAC is likely to be higher than that of IB form. This difference in mobility of IB and RB forms can be explained by the fact, that IB form of OAC is localized within the densely packed structural regions of amorphous phase of PA-6, which are characterized by a high level of intermolecular interaction. The RB form of OAC occupies loosely packed structural regions of amorphous phase with lower energetics of intermolecular interaction. As a result, molecular mobility of RB form appears at lower temperatures as compared with IB form. From this standpoint, in the temperature range from 373-383 to 413-423 K, the molecular mobility of RB form of OAC is rather high, and this form of OAC appears to plasticize the polymer, thus decreasing the elastic modulus. On contrary, molecular mobility of IB form of OAC is still frozen. As a result, this form of OAC exhibit an antiplasticizing action on PA-6. In other words, the elastic properties of PA-6/OAC samples are controlled by the interplay between the plasticizing action of RB form and the antiplasticizing action of IB form. The resulting effect appears to be decided by the ratio between IB and RB forms in the polymer matrix or fractional content of IB form in PA-6.

From this standpoint, the existence of the same T_I for all Samples I studied (Fig. 3) is well correlated to the fact, that all these samples are characterized by the same fractional content of IB form of OAC in polymer - 0.32 ± 0.02 (Fig. 2). It is of particular interest to study the effect of accumulation of IB form in PA-6 on dynamic mechanical properties of polymer. As was mentioned above, for Samples II, increasing treatment time of PA-6 films in aqueous OAC solutions results in increasing fractional content of IB form of OAC in polymer (Fig. 2). As follows from Fig. 5, with increasing fractional content of IB form in PA-6 T_I shifts towards higher temperatures.

Hence, this experimental evidence unequivocally demonstrates the influence of the molecular distribution of OAC between RB and IB forms within polymer matrix on the mechanical properties of PA-6/OAC samples. This behavior is likely to be associated with the difference in temperature-induced molecular mobility of IB and RB forms of OAC because of their localization within different structural regions in polymer amorphous phase. For detailed discussion of this phenomenon, let us consider generalized data concerning the physical and mechanical behavior (T_g, T_I, and E') of PA-6/OAC samples.

For Samples I, II, and III, Fig. 6 shows the plots of T_g and T_I versus OAC content in PA-6. This allows one to follow the behavior of these parameters of PA-6/OAC samples which are characterized by the different fractional content of IB form of OAC at fixed net content of OAC in polymer. As might be seen, the above parameters are not controlled by the net content of OAC in polymer, but by fractional content of IB form of OAC in PA-6. For Samples I which are characterized by the same fractional content of IB form, the dependence of T_g on OAC content is very weak in the range of OAC content from 0.04 to 0.5 mol/kg (Fig. 6A, curve 1). For these samples, T_I appears to be independent on OAC content (Fig. 6B, curve 1). For Samples II, with increasing periods of treatment time with aqueous OAC solutions fractional content of IB form of OAC increases, and, as a result, both T_g and T_I increases (Fig. 6, curves 2 and 3). For Samples III, removal of RB form from PA-6/OAC samples results in a dramatic growth in both T_g and T_I (Fig. 6, curves 4 and 5).

For Samples I, II, and III, Fig. 7 shows the dependence of E' on content of OAC in PA-6 at 363 K. At this temperature, a marked difference between the influence of IB and RB forms of OAC on the elastic properties of PA-6 is observed. When passing from Samples I to Samples III (transition from curves 1 to curves 4 and 5, Fig. 7), increasing fractional content of IB form of OAC in PA-6/OAC samples is responsible for the fact, that, at the same net OAC content in PA-6, for Samples III, the values of storage moduli exceed those for Samples I. For Samples II, with increasing fractional content of IB form storage moduli increase (Fig. 7, curves 2 and 3). As was mentioned above, this phenomenon may be attributed to the interplay between plasticizing action of RB form of OAC on mechanical properties of PA-6 and antiplasticizing action of IB form.

Note, that at 293 K, for Samples I, II, and III, concentration dependences of E' fit one curve. In other words, there is no difference in influence of both IB and RB forms of OAC on elastic properties of PA-6, and elastic moduli are controlled only by the net content of OAC in polymer. This experimental evidence can be explained by the fact, that, at this temperature, molecular mobility of both IB and RB forms is inhibited. As a result, both these forms have an antiplasticizing action on polymer. Similar behavior is observed at 413 K when all samples are in rubbery state. In this temperature region, molecular mobility of both OAC forms is enhanced, and their plasticizing action of PA-6 is well pronounced.

Hence, a marked difference between the influence of IB and RB forms of OAC on mechanical properties of PA-6/OAC samples is observed in the temperature region associated with the transition from glassy to rubbery state.

To investigate the effect of distribution of OAC between IB and RB forms on physical and mechanical behavior of oriented PA-6 samples, temperature-induced shrinkage of oriented polymer samples was studied (Fig. 8). For oriented initial PA-6 films (curves 1), at temperatures below melting point T_m (≈500K), temperature-induced shrinkage, which is characterized by decreasing residual deformation ε_{res}, is likely to be controlled by the relaxation of deformation within amorphous regions of polymer. With increasing temperature close to T_m residual deformation dramatically falls down because of crystallites melting.

At lower temperatures, introduction of OAC into oriented PA-6 films is responsible for suppression of shrinkage of oriented PA-6/OAC samples (curves 2). As temperature increases the shrinkage of oriented PA-6/OAC samples becomes higher as compared with initial polymer. In other words, for initial oriented PA-6 and oriented PA-6 modified with OAC, the inversion in shrinkage is observed. As for inversion in elastic properties of isotropic PA-6 films and PA-6/OAC samples, the inversion in shrinkage of oriented polymer samples may be explained by molecular mobility of OAC in polymer matrix. With increasing temperature molecular mobility of OAC in PA-6 increases, and decreasing energy of intermolecular interaction between polymer macrochains takes place. As a result, for PA-6/OAC oriented samples, temperature-induced shrinkage is higher as compared with initial PA-6.

For oriented PA-6 samples treated with aqueous OAC solution with concentration of 0.07 mol/l, at temperatures above 450 K, removal of RB form of OAC is accompanied with suppression in shrinkage as compared with initial PA-6 and PA-6 containing both IB and RB forms of OAC (Fig. 8A). At lower temperatures, for oriented PA-6 sample containing only IB form of OAC (curve 3), temperature-induced shrinkage is higher than that for PA-6 sample containing both IB and RB forms of OAC (curve 2). Note, that for oriented PA-6 samples treated with aqueous OAC solution with concentration of 0.007 mol/l, removal of RB form of OAC results in suppression in shrinkage within the whole temperature range not above melting point (Fig. 8B). Hence, for oriented PA-6 films modified with OAC, their temperature-induced shrinkage is controlled by distribution of OAC between RB and IB forms. As for isotropic PA-6/OAC samples, the influence of distribution of OAC on physical and mechanical behavior of oriented PA-6/OAC samples is decided by difference in temperature-induced molecular mobility of the above forms of OAC within polymer matrix, and, as a result, by opposite action of these forms on polymer properties in certain temperature intervals.

To gain a deeper insight into the influence of distribution of OAC between IB and RB forms on physical and mechanical behavior of PA-6/OAC samples, let us discuss the data on water sorption by the samples of the initial PA-6 and PA-6 containing OAC.

For Samples I and III, Fig. 9 shows isotherms of water sorption at 293 K by the amorphous regions of the samples studied. As can be seen, the values of water content of PA-6 and Sample III containing only IB form of OAC (curves 1 and 2, respectively) quantitatively coincide. However, for sample I with the same net content of OAC (curve 3), a marked decrease in water content is observed. In other words, the sorptional properties of PA-6/OAC samples are controlled by the fractional content of IB form of OAC in the polymer. This experimental evidence suggests that IB form of OAC is distributed in the densely packed domains of the amorphous phase of the semicrystalline polymer, which are inaccessible for water molecules in a whole range of relative humidities. The RB form of OAC occupies disordered regions of the PA-6 amorphous phase, which are responsible for water sorption of PA-6. The introduction of OAC molecules into the above disordered regions is accompanied by physical cross-linking of macromolecules via formation of strong hydrogen bonds between phenolic hydroxyls of OAC and amide groups of PA-6. This gives rise to a drop in sorption capacity of such material as compared with the initial polymer.

The studies of elastic properties of the PA-6/OAC samples in a humid atmosphere also support a complicated pattern of influence of OAC on the mechanical properties of the polymer. Figure 10 shows the dependences of elastic modulus E_0 of PA-6/OAC samples versus OAC content at $p/p_0 = 0.47$ and 0.66. Curves 1 and 2 correspond to the Samples I for which with increasing net content of OAC fractional content of IB form remains constant - 0.32 ± 0.02. Curves 3, 4, and 5 are associated with Samples II. These samples show an increase in fractional content of IB form of OAC with increasing the net content of OAC in polymer. Curves 6 and 7 demonstrate the variations in elastic properties of the PA-6/OAC samples upon desorption of RB form of OAC (transition from Samples I to Samples III). In this case, with decreasing net content of OAC in polymer fractional content of IB form of OAC increases.

As might be seen, elastic modulus of the PA-6/OAC samples at given values of p/p_0 is determined not only by the net content of OAC in the polymer, but also by distribution of OAC between IB and RB forms. An increase in net content of OAC in the polymer amorphous regions is accompanied by increase in elastic modulus of the PA-6/OAC samples (Fig. 6, curves 1 - 5). However, the growth in E_0 is more pronounced, when an increase in the net OAC content is associated with increase in the fractional content of IB form (curves 3 - 5). Note that, in a wide range of OAC concentrations, the removal of RB form of OAC from the PA-6/OAC samples (curves 6 and 7) has no influence on elastic modulus of the polymer.

Let us emphasize, that the PA-6/OAC samples containing both IB and RB forms of OAC (Samples I and II) show a dramatic drop in water content as compared with initial PA-6 within the whole range of relative humidities. In this case, a decreased water sorption is controlled by the net content of OAC in the polymer. In other words, water sorption of the Samples I and II with the same net content of OAC (curves 1 - 5) is the same, despite a certain difference in the ratio between IB and RB forms of OAC. Desorption of RB form of OAC from PA-6 is accompanied by a sharp increase in water sorption. Note that water sorption of the Samples III containing only IB form of OAC is similar to that of initial PA-6 (Fig. 9, curves 1 and 2). Nevertheless, despite a high content of plasticizer (water) in the polymer, the elastic modulus of the samples is much higher than the corresponding values for initial PA-6 and Samples I and II containing both IB and RB forms of OAC.

The data presented in Fig. 10 allows the following general conclusion. For the PA-6/OAC samples with the same net content of OAC, the higher modulus is observed for the samples with higher fractional content of IB form of OAC. Distribution of OAC in the polymer matrix between IB and RB forms controls not only the absolute values of E_0 of the PA-6/OAC samples, but also sensitivity of these characteristics to the plasticizing action of water. Increase in the fractional content of IB form of OAC is accompanied by a concomitant growth in E_0, even though water sorption of the samples remains unchanged (curves 2 - 5) or drastically increases (curves 1, 6 and 7). The influence of IB form of OAC accumulated in the densely packed domains of the amorphous phase on elastic properties of the polymer proves the important role of these structural regions for the development of physico-mechanical properties of semicrystalline polymers.

Let us emphasize, that the above phenomena were observed only in the intermediate range of relative humidities, i.e., at 0.47 and 0.66. At $p/p_0 = 0$ or 1, the PA-6/OAC samples are in glassy or rubbery state, respectively. In this case, the elastic properties of the samples do not recognize distribution of OAC between IB and RB forms. They depend only on the net content of OAC in the PA-6/OAC samples. In other words, the decisive role of the ratio between IB and RB forms of OAC is well pronounced only in the range of relative humidities associated with transition from glassy to rubbery state.

CONCLUSION

Introduction of OAC in amorphous regions of PA-6 is associated with the distribution of OAC between two different forms reversibly and irreversibly bound with polymer matrix. The ratio between the above forms of OAC in polymer can be controlled either by varying the regimes of treatment of PA-6 samples in aqueous OAC solutions or by subsequent treatment of PA-6/OAC samples in water. Generally speaking, this ratio is responsible for the physical and mechanical behavior of PA-6 samples modified with OAC: glass transition temperature, temperature range of reinforcing action of OAC on PA-6 mechanical properties, water sorption, and elastic properties. Note, that elastic properties of PA-6 containing OAC are controlled by distribution of OAC between the above forms only in the case when polymer undergoes transition of its amorphous phase from glassy to rubbery state, regardless of whether this is realized: by increasing the concentration of plasticizer (water) or by increasing the temperature. This behavior is likely to be associated with the localization of reversibly and irreversibly bound forms of OAC within different structural sublevels in polymer amorphous phase.

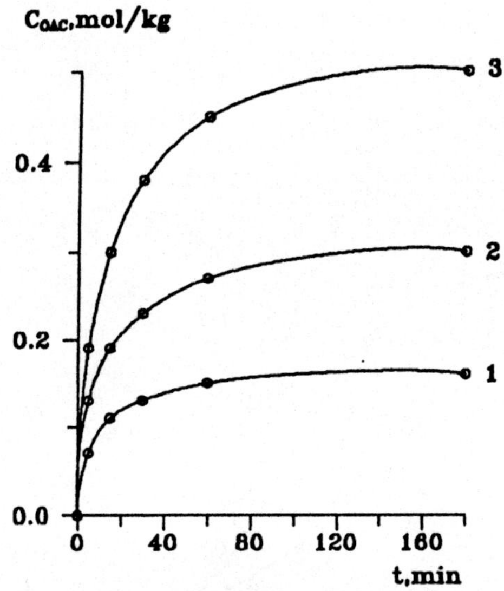

Fig. 1. Kinetic curves of OAC sorption by PA-6 films from aqueous OAC solutions with concentrations 0.036 (1), 0.072 (2), and 0.14 mol/l (3).

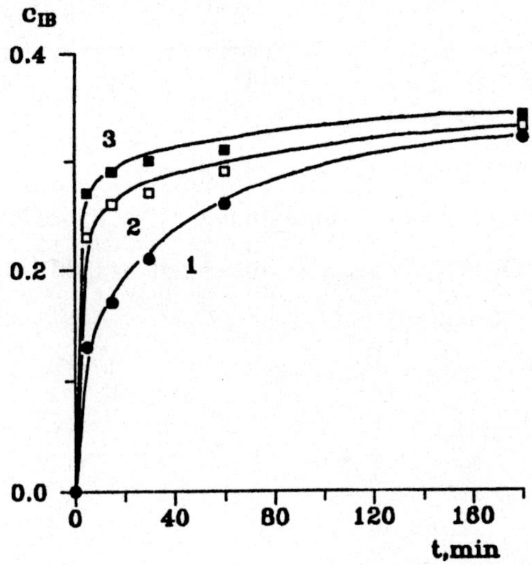

Fig. 2. Fractional content C_{IB} of IB form of OAC in PA-6 films versus time of treatment of PA-6 with aqueous OAC solutions with concentrations 0.036 (1), 0.072 (2), and 0.14 mol/l (3).

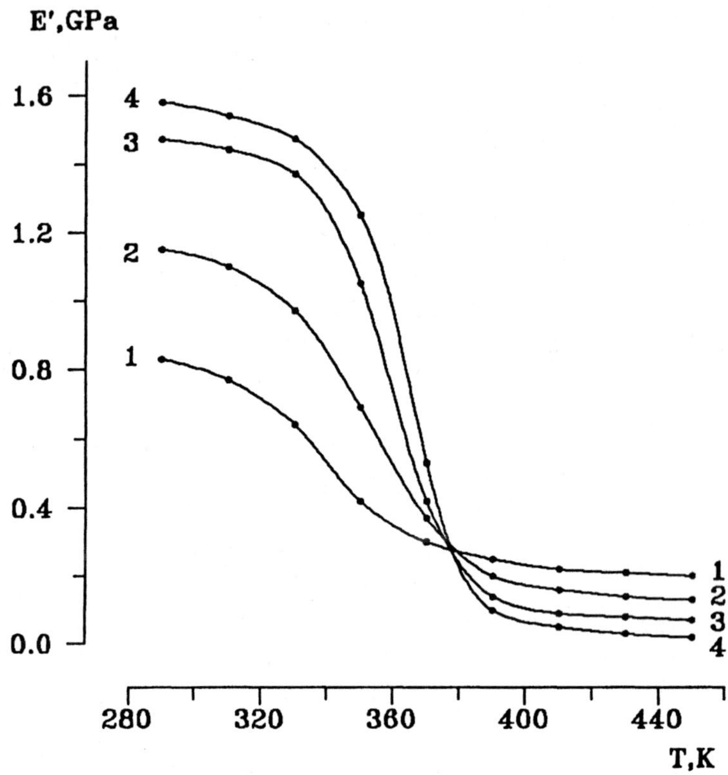

Fig. 3. Temperature dependences of storage moduli E' of initial PA-6 (1) and PA-6 treated with aqueous OAC solutions with concentrations 0.007 (2), 0.072 (3), and 0.14 mol/l (4) for 3 h (Samples I).

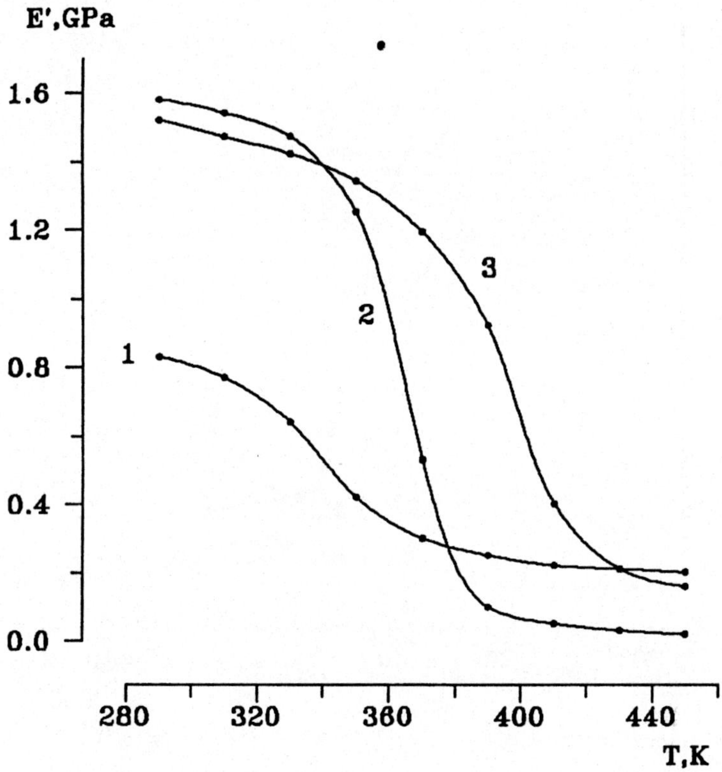

Fig. 4. Temperature dependences of storage moduli E' of initial PA-6 (1), PA-6 treated with aqueous OAC solution with concentration 0.14 mol/l for 3 h (Sample I) (2), and, then, with boiling water for 6 h (Sample III) (3).

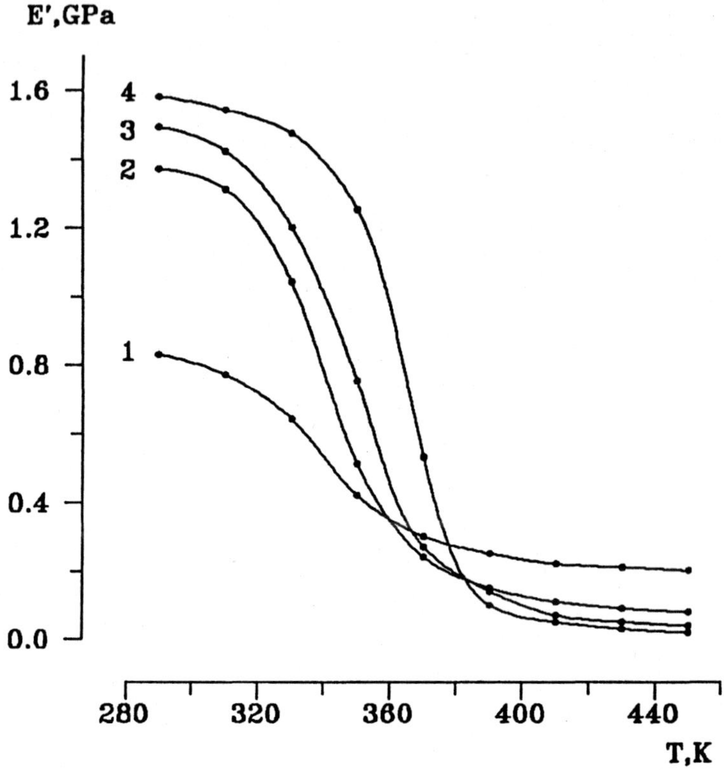

Fig. 5. Temperature dependences of storage moduli E' of initial PA-6 (1) and PA-6 treated with aqueous OAC solution with concentration 0.14 mol/l for 5 (2), 30 (3), and 180 min (4) (Samples II).

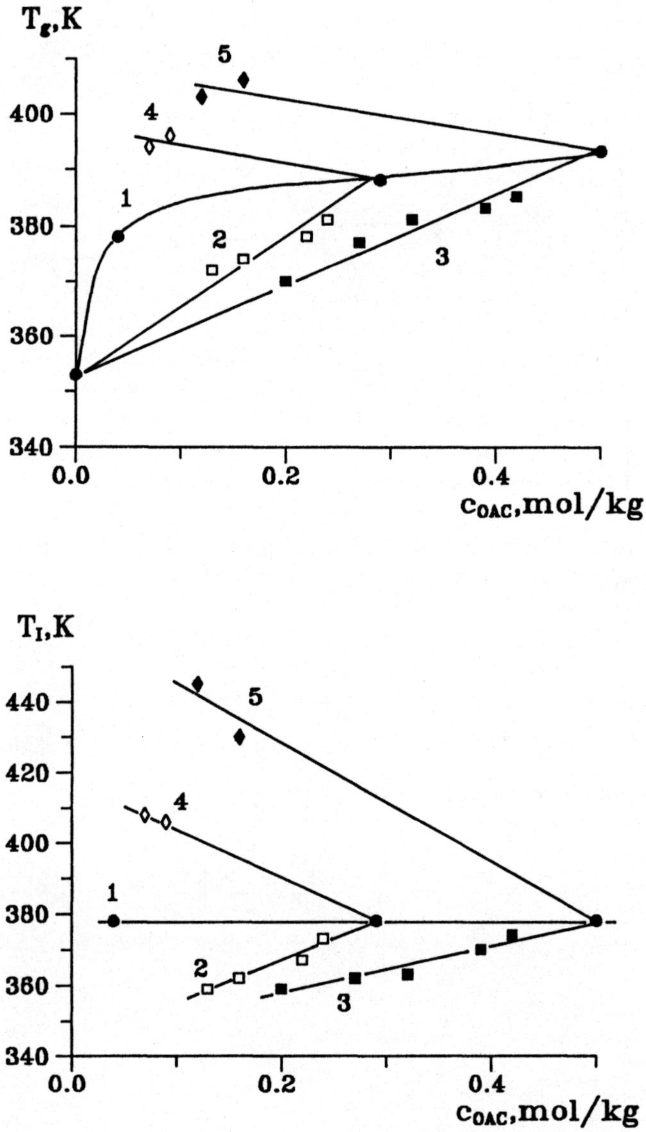

Fig. 6. Glass transition temperature T_g (A) and inversion point of elastic modulus T_I (B) versus OAC content in PA-6. Samples I (1), Samples II (2 and 3), and Samples III (4 and 5).

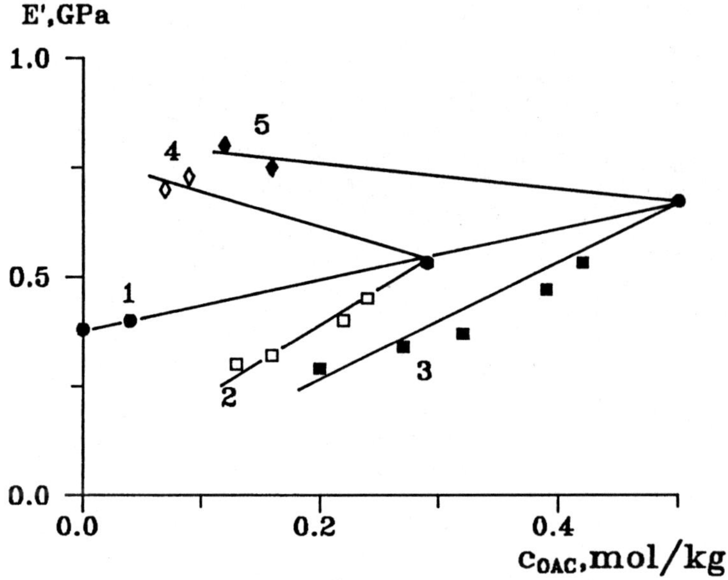

Fig. 7. Storage moduli E' versus OAC content in PA-6 at 363 K. Samples I (1), Samples II (2 and 3), and Samples III (4 and 5).

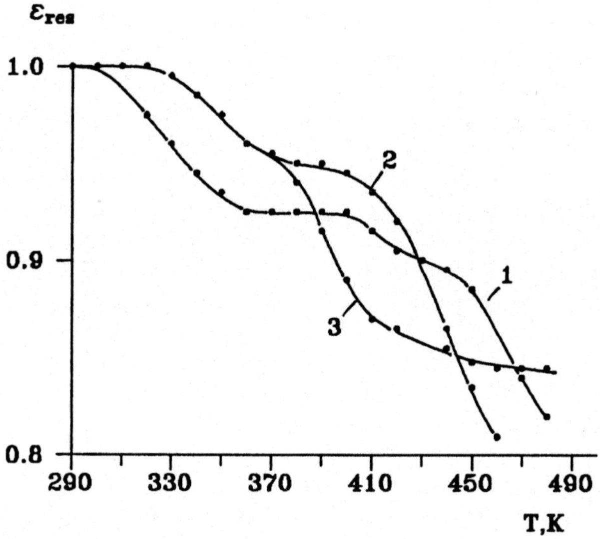

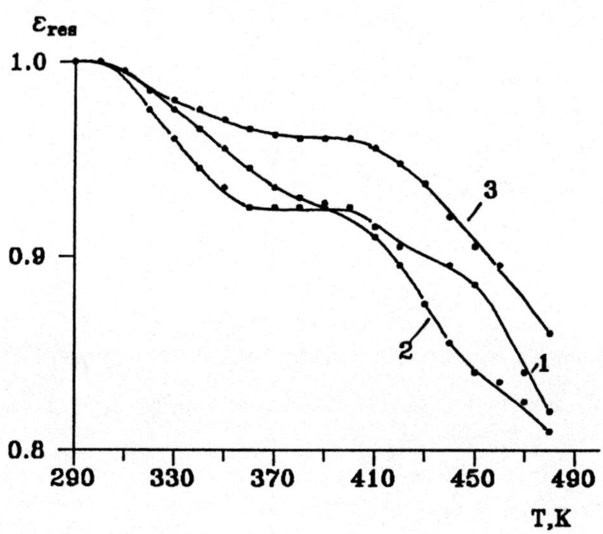

Fig. 8. Temperature dependences of relative residual deformation ε_{res} of initial oriented PA-6 (1), oriented PA-6 treated with aqueous OAC solutions with concentrations 0.07 (A) and 0.007 mol/l (B) for 3 h (2), and, then, with boiling water for 6 h (3).

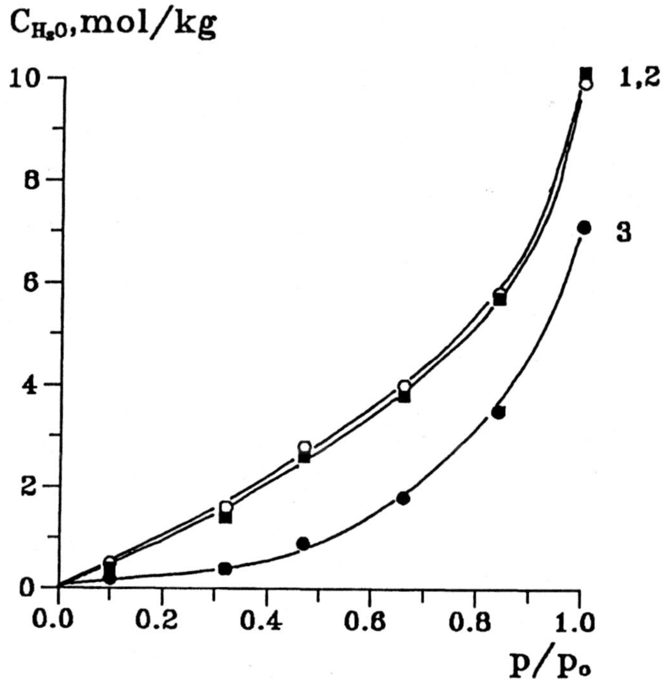

Fig. 9. Isotherms of water sorption (293 K) by amorphous regions of initial PA-6 (1) and PA-6/OAC samples with OAC content 0.14 mol/kg (2, 3). 2 - Sample III, 3 - Sample I.

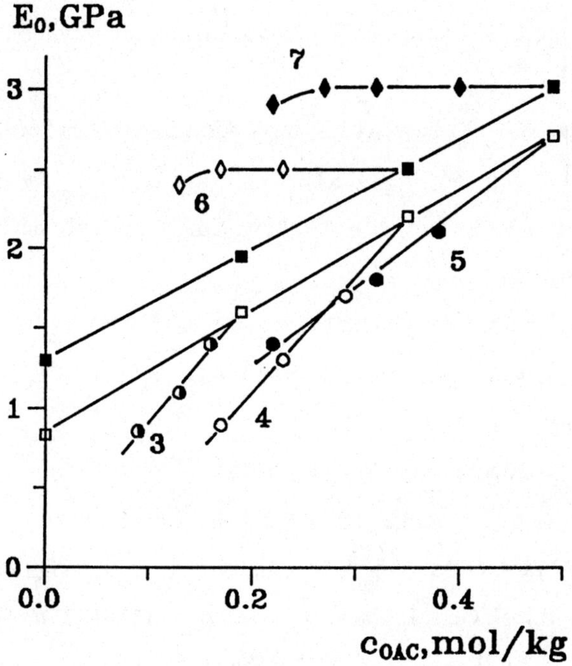

Fig. 10. Elastic moduli of PA-6/OAC samples versus OAC content at p/p_0 0.47 (1, 6, 7) and 0.66 (2 - 5). Samples I (1, 2), Samples II (3 - 5), Samples III (6, 7).

REFERENCES

1. M. Lewin, H. Guttman and Y. Naor, J. Macromol. Sci., A, 1367 (1988).

2. A.Ya. Polishchuk, L.A. Zimina, N.N. Madyuskin and G.E. Zaikov, Polym. Sci., A, 35, 80 (1993).

3. M.S. Arzhakov, A.V. Volkov, A.L. Volynskii and N.F. Bakeev, Makromol. Chem., Macromol. Symp., 44, 165 (1991).

4. M.S. Arzhakov, N.V. Red'ko, A.V. Volkov, A.L. Volynskii and N.F. Bakeev, Polym. Sci., A, 36, 929 (1994).

5. M.S. Arzhakov, Intern. J. Polymeric Mater., 28, 103 (1995).

6. M.S. Arzhakov, E.V. Nechaeva, N.V. Red'ko, A.V. Volkov, A.L. Volynskii and N.F. Bakeev, Polym. Sci., A, 35, 62 (1996).

7. D.C. Prevorsec, J. Polym. Sci., 32, 434 (1971).

8. D.C. Prevorsec, G.A. Tirpac, P.J. Harget and H.C.Reimschuessel, J. Macromol. Sci., B, 9, 733 (1974). 9. G.S.J. Yeh, J. Macromol. Sci., B, 6, 465 (1972).

10. I.J. Klement and P.H. Geil, J. Macromol. Sci., b, 5, 505 (1971).

11. S.A. Arzhakov, N.F. Bakeev and V.A. Kabanov, Vysokomol. Soedin., A, 15, 1154 (1973) (in Russian).

12. G.V. Kozlov, D.S. Sanditov and V.D. Serdyuk, Polym. Sci., A, 35, 1742 (1993).

13. I.I. Perepechko and S.S. Yakovenko, Vysokomol. Soedin., A, 23, 1166 (1981) (in Russian).

14. N.S. Murphy, M. Stamm, J.P. Sibilia and S. Krimm, Macromolecules, 22, 1261 (1989).

15. N.S. Murphy and W.J. Orts, J. Polym. Sci., Polym. Phys., 32, 2695 (1994).

16. M.S. Arzhakov, Cand. Sci. (Chem.) Dissertation, Moscow State University, Moscow, 1989.

17. J. Guthrie and C.C. Cook, Polymer, 22, 1439 (1981).

18. M. Topic, A. Mogus-Milankovic and Z. Katovic, Polymer, 28, 33 (1987).

19. M.S. Arzhakov, A.V. Volkov, A.L. Volynskii and N.F. Bakeev, Polym. Sci., A, 35, 521 (1993).

20. M.S. Arzhakov, A.V. Volkov, A.L. Volynskii and N.F. Bakeev, Intern. J. Polymeric Mater., 25, 89 (1993).

Subject Index

Subject Index